BEYOND BASIC STATISTICS

BEYOND BASIC STATISTICS

Tips, Tricks, and Techniques Every Data Analyst Should Know

KRISTIN H. JARMAN

Published by John Wiley & Sons, Inc., Hoboken, New Jersey
Published simultaneously in Canada

For general information on our other products and services or for technical support, please contact our Customer Care Department within the United States at (800) 762-2974, outside the United States at (317) 572-3993 or fax (317) 572-4002.

Wiley also publishes its books in a variety of electronic formats. Some content that appears in print may not be available in electronic formats. For more information about Wiley products, visit our web site at www.wiley.com.

Library of Congress Cataloging-in-Publication Data:

Jarman, Kristin H.
 Beyond basic statistics: tips, tricks, and techniques every data analyst should know / K.H. Jarman.
  pages   cm
 Includes index.
 ISBN 978-1-118-85611-6 (pbk.)
1. Mathematical statistics–Popular works.   I. Title.
 QA276.J37 2015
 001.4′22–dc23

2014047952

Set in 10.5/13pt Times by SPi Publisher Services, Pondicherry, India

Printed in the United States of America

10 9 8 7 6 5 4 3 2 1

1   2015

To Anna and Sarah

When something doesn't feel right, it probably isn't.

CONTENTS

Preface   ix

1   Introduction: It Seemed like the Right Thing to Do
at the Time   1

2   The Type A Diet: Sampling Strategies to Eliminate
Confounding and Reduce Your Waistline   9

3   Conservatives, Liberals, and Other Political Pawns:
How to Gain Power and Influence with Sample
Size Calculations   31

4   Bunco, Bricks, and Marked Cards: Chi-Squared
Tests and How to Beat a Cheater   47

5   Why it Pays to be a Stable Master: Sumo Wrestlers
and Other Robust Statistics   69

6   Five-Hour Marriages: Continuous Distributions, Tests for
Normality, and Juicy Hollywood Scandals   91

7   Believe It or Don't: Using Outlier Detection to Find the
Weirdest of the Weird   109

8 The Battle of the Movie Monsters, Round Two:
Ramping up Hypothesis Tests with Nonparametric Statistics   123

9 Models, Murphy's Law, and Public Humiliation:
Regression Rules to Live By   139

Appendix A   Critical Values for the Standard Normal
Distribution   163

Appendix B   Critical Values for the T-Distribution   165

Appendix C   Critical Values for the Chi-Squared Distribution   167

Appendix D   Critical Values for Grubbs' Test   169

Appendix E   Critical Values for Wilcoxson Signed
Rank Test: Small Sample Sizes   171

Glossary   173
Index   185

PREFACE

I've had my share of mistakes: spilled coffee, insensitive remarks, and red socks thrown into a load of white laundry. These are daily occurrences in my life. But it isn't these little, private mishaps that haunt me. It's the big ones, the data analysis disasters, the public humiliations resulting from my own carelessness, mistakes that only reveal themselves when I'm standing in front of a room full of important people, declaring the brilliance of my statistical conclusions to the world.

Fortunately, these humiliations appear much more often in my dreams than they do in real life. When they do happen, however, they hit me when I least expect them, when I'm rushed, or when I'm overconfident in my results. All of them are accidental. I certainly never mean to misinform, but when you analyze as much data as I do, small mistakes are bound to happen every now and then.

This book highlights some of the well-known shortcomings of basic statistics, shortcomings that can, if ignored, lead to false conclusions. It provides tips and tricks to help you spot problem areas in your data analysis and covers techniques to help you overcome them. If, somewhere within the chapters of this book, you find information that prevents you from experiencing your own statistical humiliation, then exposing my own embarrassment will have been worth it.

KRISTIN H. JARMAN

1

INTRODUCTION: IT SEEMED LIKE THE RIGHT THING TO DO AT THE TIME

As a seasoned statistical scientist, I like to think I'm invincible when it comes to drawing reliable conclusions from data. I'm not, of course. Nobody is. Even the world's best data analysts make mistakes now and then. This is what makes us human.

Just recently, for example, I was humbled by the simplest of all statistical techniques: the confidence interval. I was working with a government panel, helping them to establish criteria for certifying devices that detect certain toxic substances. (Smoke detectors, for example, are certified so you know they're reliable; in other words, they're likely to sound an alarm when there's smoke, and keep quiet when there isn't). The committee members wanted to know how many samples to test in order to reach a certain confidence level on the probability of detection, the probability that, given the toxin is present, the device will actually sound an alarm.

No problem, I thought.

Back in my office, I grabbed a basic statistics book, pulled out the formula for a confidence interval of a proportion (or probability), and went to work. I began calculating the confidence bounds on the probability of detection for different testing scenarios, preparing recommendations as I went along. It wasn't until sometime later I realized all my calculations were wrong.

Beyond Basic Statistics: Tips, Tricks, and Techniques Every Data Analyst Should Know, First Edition. Kristin H. Jarman.
© 2015 John Wiley & Sons, Inc. Published 2015 by John Wiley & Sons, Inc.

Well, not wrong, the formulas and numbers were correct. But they didn't really fit my problem. When I started the calculations, I'd neglected one small but important detail. The detection probability for the devices being tested is typically very high, say 0.95 or higher. The basic confidence interval for a proportion p uses a normal approximation, which only applies when $Np > 5$ and $N(1-p) > 5$. Since I was limited to relatively small sample sizes of $N = 80$ or less, at best I had $N(1-p) = 80 \times 0.05 = 4$. Not large enough for the standard confidence interval to apply.

This happens more than I care to admit, that I embark on a data analysis using the world's most common statistical techniques, only to realize that my data don't work with the tools I'm using. Maybe the data don't fit the nice, bell-shaped distribution required by most popular methods. Maybe there are extreme values that could skew my results. But whatever the problem, I know that if I don't address it or at least acknowledge the impact it might have on my results, I will be sorry in the end.

This book takes you beyond the basic statistical techniques, showing you how to uncover and deal with those less-than-perfect datasets that occur in the real world. In the following chapters, you'll be introduced to methods for finding outliers, determining if a sample conforms to a normal distribution, and testing hypotheses when your data aren't normal. You'll learn popular strategies for designing experimental studies and performing regression with multiple variables and polynomial functions. And you'll find many tips and tricks for dealing with difficult data.

WHEN GOOD STATISTICS GO BAD: COMMON MISTAKES AND THE IMPACT THEY HAVE

There are many ways good statistics can go wrong and many more ways they can impact a data analyst's life. But in my experience, the vast majority of these mishaps are caused by just a few relatively common mistakes:

- Answering the wrong question
- Gathering the wrong data
- Using the wrong statistical technique
- Misinterpreting the results

Anyone who deals with a lot of data commits at least one of these errors from time to time. In my most recent incident, where I was slapped down by a simple confidence interval, I was clearly applying the wrong technique. Thankfully, this error only cost me a little time and it was easily fixed.

In Chapter 9, I'll share another one of my statistical humiliations, a situation where I misinterpreted the results of an analysis, a mistake that could've ruined my reputation and cost my employer millions of dollars.

This book introduces many statistical techniques designed to keep you from making these four common errors. Chapters 2 and 3 focus on designing studies based on your research goals. Chapters 5–9 introduce statistical techniques that can help you select the right analysis for a particular problem. In all of the chapters, the emphasis lies not on the mathematics of statistics but on how and when to use different techniques so you can avoid making costly mistakes.

STATISTICS 101: CONCEPTS YOU SHOULD KNOW BEFORE READING THIS BOOK

The techniques taught in most introductory statistics classes are built on a relatively small number of concepts, things like the sample mean and the normal distribution. But not-so-basic techniques are built on them, too. Before you dive too deeply into the world of data analysis, it's important to have a working knowledge of a handful of concepts. Here are the ones you'll need to get the most out of this book. For a detailed introduction to these topics, see a basic statistics textbook such as the companion to this book, *The Art of Data Analysis: How to Answer Almost Any Question Using Basic Statistics* by yours truly.

Probability Theory

Statistics and data analysis rely heavily on mathematical probability. Mathematical probability is concerned with describing randomness, and all of the functions and complex formulas you see in a statistics book were derived from this branch of mathematics. To understand the techniques presented in this book, you should be familiar with the following topics from probability.

Random Variables and Probability Distributions   A **random variable** represents the outcome of a random experiment. Typically denoted by a capital letter such as X or Y, a random variable is similar to a variable x or y from algebra. Where the variable x or y represents some as yet unsolved value in an algebraic equation, the variable X or Y represents some as-yet-undetermined outcome of a random experiment. For example, on a coin toss, with possible outcomes *heads* and *tails*, you could define a random variable $X = 0$ for *tails* and $X = 1$ for *heads*. This value of X is undetermined until the experiment is complete.

A **probability distribution** is a mathematical formula for assigning probabilities to the outcomes of a random experiment. Many different probability distributions have been developed over the years, and these can be used to assign probabilities in almost any random experiment you can imagine. Whether or not you'll win the lottery, how many times your new car will break down in the first year, the amount of radioactivity you'll absorb while scooping out your cat's litter box, all of these events have a probability distribution associated with them.

Expected Values and Parameters of a Distribution   A random variable is uncertain. You don't know exactly what value it will take until the experiment is over. You can, however, make predictions. The **expected value** is just that: a prediction as to what value a random variable will take on. The two most common expected values are the mean and variance. The mean predicts the value of the random variable, and the variance predicts the likely deviation from the mean. The **parameters** of a distribution are values that specify the exact behavior of a random variable. Every probability distribution has at least one parameter associated with it. The most common parameters are also expected values: in particular, the mean and variance.

Statistics

Statistics is the application of probability to real data. Where probability is concerned with describing the mathematical properties of random variables, statistics is concerned with estimating or predicting mathematical properties from a set of observations. Here are the basic concepts used in this book.

Population vs. Sample   In any study, the goal is to learn something about a **population**, the collection of all people, places, or things you are interested in. It's usually too costly or too time-consuming to collect data from the entire population, so you typically must rely on a **sample**, a carefully selected subset of the population.

Parameter vs. Estimate   A parameter is a value that characterizes a probability distribution, or a population. An estimate is a value calculated from a dataset that estimates the corresponding population parameter. For example, think of the population mean and sample mean, or average. The population mean is a parameter, the true (often unknown) center of the population. The sample mean is an estimate, an educated guess as to what the population mean might be.

Discrete vs. Continuous Data   Any data collection exercise produces one or more outcomes, and these outcomes—called observations, measurements, or data—can be either discrete or continuous. **Discrete observations** are whole numbers, counts, or categories, in other words, anything that can be easily listed. For example, the outcome of one roll of a six-sided die is discrete. **Continuous observations**, on the other hand, cannot be listed. Real numbers are continuous. If you choose any two real numbers, no matter which two you choose, there's always some number in between them. Different statistical techniques are often applied to discrete and continuous data.

Descriptive Statistics   **Descriptive statistics** are estimates for the center location, shape, texture, and other properties of a population. Descriptive statistics are the foundation of data analysis. They're used to describe a sample, construct margins of error, compare two datasets, find relationships between variables, and just about anything else you might want to do with your data. The two most common descriptive statistics are the sample mean (average) and standard deviation.

The **average**, or **sample mean**, describes center location of a sample. Calculated as the sum of all your data values divided by the number of data values in the dataset, the average is the arithmetic center of a set of observations. The standard deviation measures the spread of a set of observations. The **standard deviation** is the average deviation, or variation, of all the values around the center location.

Sample Statistics and Sample Distributions   A **sample statistic** is calculated from a dataset. It's a value with certain statistical properties that can be used to construct confidence intervals and perform hypothesis tests. A z-statistic is an example of a sample statistic. A **sample distribution** is a probability distribution for a sample statistic. Critical thresholds and p-values used in confidence intervals and hypothesis tests are calculated from sample distributions. Examples of such distributions include the z-distribution and the t-distribution.

Confidence Intervals   A **confidence interval**, or **margin of error**, is a measure of confidence in a descriptive statistic, most commonly the sample mean. Confidence intervals are typically reported as a mean value plus or minus some margin of error, say 8 ± 2 or as a corresponding range, such as $(6, 10)$.

Hypothesis Tests   A **hypothesis test** uses data to compare competing claims about a population in order to determine which claim is most likely. There are typically two hypotheses being compared: H_0 and H_A. H_0 is called the **null**

hypothesis. It's the fall-back position. It's what you're automatically assuming to be true. H_A is the **alternative hypothesis**. This is the claim you accept as true only if you have enough evidence in the data to reject H_0.

Hypothesis tests are performed by comparing a test statistic to a critical threshold. The **test statistic** is a sample statistic, a value calculated from the data. This value carries evidence for or against H_0. The **critical threshold** is a value calculated from a sample distribution and the **significance level**, or probability of falsely rejecting H_0. You compare the test statistic to this threshold in order to decide whether to accept that H_0 is true, or reject H_0 in favor of H_A.

Alternatively, you can use the test statistic to calculate a *p*-value, a probability for the evidence under the null hypothesis, and compare it to the significance level of the test. If the *p*-value is smaller than the significance level, then H_0 is rejected.

In general, hypothesis tests are either one-sided or two-sided. A **one-sided hypothesis test** looks for deviations from the null hypothesis in one direction only, for example, when testing if the mean of a population is zero or *greater than* zero. A **two-sided hypothesis test** looks for deviations in both directions, as in testing whether the mean of a population is zero or *not equal* to zero. One-sided and two-sided hypothesis tests often have the same test statistic, but to achieve the same significance level, they typically end up using use different critical thresholds.

Linear Regression   **Linear regression** is a common modeling technique for predicting the value of a dependent variable Y from a set of independent X variables. In linear regression, a line is used to describe the relationship between the Xs and Y. **Simple linear regression** is linear regression with a single X and Y variable.

TIPS, TRICKS, AND TECHNIQUES: A ROADMAP OF WHAT FOLLOWS

Each chapter in this book begins by asking a specific question and reviewing the basic statistics approach to answering it. Common problems that can derail the basic approach are presented, followed by a discussion of methods for overcoming them. Along the way, tips and tricks are introduced, taking you beyond the techniques themselves into the real-world application of them. In most cases, the chapter wraps up with a case study that pulls the different concepts together and answers the question posed at the beginning.

Where basic statistics and a little algebra can be used to explain a technique, the mathematical details are included. However, in several cases, the mathematics goes beyond the basics, requiring more advanced tools such as calculus and linear algebra. In those cases, rather than presenting the mathematical details of a method, I focus instead on the big picture, what the technique does and how to use it. With this strategy, I hope to avoid getting bogged down with the math, and keeping emphasis on the application of the methods to real world situations.

A final note regarding data analysis software. There are many statistical software packages out there, and every data analyst has his or her personal favorite. Most hard core data analysts eventually migrate to powerful tools such as R, Matlab, or SAS. Most of these have an interface, much like a programming language, that allows you to tailor your data analyses in almost any way you'd like. But there are less sophisticated tools that are more user friendly and have a wide variety of useful techniques built into them. These tools are good for beginners and those who fear programming, and if you don't already have a favorite data analysis software, I urge you to search the Internet for one. For this book, I downloaded a popular Excel add-in. Every analysis you'll find in this book was done entirely in Excel, or with the help of this inexpensive add-in.

BIBLIOGRAPHY

eHow.com. How to Avoid Common Errors in Statistics. Available at http://www.ehow.com/how_2294991_avoid-common-errors-statistics.html. Accessed November 1, 2013.

Good PI, Hardin JW. *Common Errors in Statistics (and How to Avoid Them)*. New York: John Wiley & Sons, Inc; 2012.

Taylor C. Common Statistics Mistakes. Available at http://statistics.about.com/od/HelpandTutorials/a/Common-Statistics-Mistakes.htm. Accessed November 1, 2013.

2

THE TYPE A DIET: SAMPLING STRATEGIES TO ELIMINATE CONFOUNDING AND REDUCE YOUR WAISTLINE

Be warned. I am not a medical doctor, nurse, dietician, or nutrition scientist. I've eaten my share of cupcakes and red meat. I've been on fad diets and dropped ten pounds only to gain them right back again. In other words, I have no real authority when it comes to any food choices, much less healthy ones. So, if you choose to follow a diet plan like the one laid out in this chapter, you do it at your own risk. And if this warning isn't enough and you still want to try it, do yourself a favor and talk to a medical professional first.

Over forty million Americans go on a diet every year. That's over forty million people looking for a way to lose weight and get healthy. And even though we all know the formula for a lean, toned body—to eat right and exercise—many of us are looking for a magic solution. Something easy to follow. Something that will keep us from binging on donuts. Something that will work fast.

In wishing for a quick and easy diet solution, I'm as guilty as anyone. I have a shelf stuffed full of diet books I've collected over the years. Every one of these books is written by an expert, somebody with a college degree who claims to have helped thousands of patients. Every one of these books claims to have the answer to long life, sexy abs, and good health. And every one of them cites numerous scientific publications to back up their claims. There's

Beyond Basic Statistics: Tips, Tricks, and Techniques Every Data Analyst Should Know,
First Edition. Kristin H. Jarman.
© 2015 John Wiley & Sons, Inc. Published 2015 by John Wiley & Sons, Inc.

only one problem. None of the books agree on much of anything. The caveman diet book claims foods like rice and potatoes cause obesity and insists we should eat meat, meat, and more meat. The happy vegetarian diet book insists meat makes us overweight and unhealthy, so we should stick to rice and potatoes. The combine-it-right diet book claims we can eat meat *and* rice *and* potatoes, so long as we don't eat them together.

How can so many experts come to such different conclusions about the foods we eat? Are they all just scam artists, slick salespeople looking to separate us from our money? And what about their so-called scientific studies? Are those faked in order to convince us to buy books and merchandise?

There are scam artists out there, to be sure. But even excluding the fraudsters, there would still be a pile of conflicting studies about health and diet. Why? Because humans are complex creatures, and studying humans is a complex job. Virtually any difficulty that can ruin an experimental plan pops up in human studies. Practical and ethical limitations. Uncontrollable variables. Confounding factors. This makes it very difficult to run an experimental study on a small number of people that can be generalized to the entire population.

Fortunately, there are tips, tricks, and techniques that can help minimize these difficulties. In this chapter, different types of experimental studies are introduced. Common mistakes are presented, along with strategies for planning an experiment—human or otherwise—that will produce the most reliable results. These strategies will be used to design a series of experiments that eliminate confounding and might even reduce your waistline.

THE BASICS OF PLANNING A STUDY

Suppose I've come up with a brand new fad diet: the Type A diet. Designed specifically for the classic Type A personality—competitive, driven, busy—this diet is just as efficient as the demographic it serves. No pesky food choices. No counting calories or carbs. Whenever you get hungry, you simply munch on one of my delicious diet bars. What could be more streamlined and simple?

In order for this weight loss plan to be successful, my diet bars need to be tasty and nutritious, and they need to promote weight loss. As a Type A personality myself, I'm not satisfied with tolerable flavor or a reasonably good nutrition profile. I want something that will pummel all those other energy and nutrition bars into the shredded cardboard they taste like. In order to beat the competition, I need to uncover their strengths and weaknesses. I need to do some research.

I'll conduct this research and design my weight loss plan through a series of studies. A **study** is a data collection exercise designed to answer some question about a group of people, places, or items. How accurately does this medical device measure blood oxygen levels? Do customers prefer a bigger smartphone with an easy-to-read screen, or a more compact device that'll fit into small pockets? Does eating thirteen grapefruit a day cause the average dieter to lose weight? All of these questions can be answered by gathering data and using statistical methods to make sense of it.

In any study, the goal is to learn something about a population, the collection of all people, places, or things you are interested in. It's usually impractical to collect data from the entire population, so you typically must rely on a sample, a carefully selected subset of the population. For example, using the entire population of humans to test whether thirteen grapefruit a day induces weight loss would be impossible. You'd have to rely on a group of test subjects to represent all humans and draw your conclusions from that group.

Every study contains both dependent and independent variables. A **dependent variable** is the outcome, the phenomenon you're studying. In the thirteen-grapefruit-a-day weight loss study, for example, the dependent variable might be the weight of test subjects taken before and after following the diet. **Independent variables** are factors you manipulate or observe in hopes of impacting the dependent variable. The number and variety of grapefruit eaten might be the independent variables in the grapefruit study. You might manipulate these things in order to determine, for example, whether a daily dose of eleven red grapefruit or fifteen yellow grapefruit promotes the most weight loss.

When planning a study, you identify the population, choose a sample, pick a dependent variable to measure, and decide how to manipulate or measure the independent variables. Sounds fairly simple, right? Unfortunately, there are many ways an experimental plan can go wrong. The next section lists some of the more common ones.

MY STATISTICAL ANALYSIS IS BRILLIANT. WHY ARE MY CONCLUSIONS SO WRONG?

There's an old saying that applies to anyone who's ever used a real-world dataset: *garbage in, garbage out*. In other words, proper data analysis is an important part of any good study, but there are no statistical techniques that can make up for bad data. If the observations you're feeding your statistical routines don't accurately represent the population or the outcome you want to measure, then the only thing statistics can do is give you false confidence in potentially inaccurate conclusions. Here are some easy-to-make mistakes when it comes to planning a study.

Answering the Wrong Question

Suppose a friend of mine suggests meal replacement shakes would be better than nutrition bars for my Type A diet. True to my perfectionist nature, I can't rest until I've either convinced myself this is true, or proven my friend wrong. I plan a study using volunteers. Half the volunteers are allowed to drink only meal replacement shakes from a popular diet plan. The other half are only allowed to eat nutrition bars. They stick to my diet for five days and come back for an evaluation at the end of it. I weigh the volunteers and ask questions about whether or not they were satisfied with the diet food they were given.

As I look over the data, a few observations jump out at me: (i) the bar-eating dieters reported being satisfied with the food more often than the shake drinkers, (ii) on average, the shake drinking group lost six pounds while their bar-eating counterparts lost only three, and (iii) several of the shake drinkers mentioned how drinking a shake too close to bed kept them up at night. I've invested a lot of time and effort planning this study and I want to get the most of it, but now I wonder: what does it mean for one diet plan to be better than another? Does an extra three-pound weight difference in the first week matter the most? Or is it better to be satisfied with the food? Will a shake that causes insomnia turn my diet plan into a flop, or will a typical conquer-the-world, Type A person appreciate the extra waking hours in the day? Unfortunately, after all my time and effort, I still have more questions than answers.

In any study, it's important to spend some time thinking about the outcome you'd like to measure and determining how it relates to the question you're trying to answer. Before I start the bar versus shake study, for example, I should think about what it means for a diet to be good. Is it only weight loss that matters? What about satisfaction with the food? Or maybe it refers to more intangible outcomes, things like energy level and overall happiness? I should also think through the different possible outcomes and what I can conclude in each case. For example, what could I conclude if one group lost more weight, but the other was more satisfied with the food? What if it was a virtual tie? Are there additional outcomes, like a dieter's energy level, that could strengthen my conclusions? By thinking through these possible scenarios beforehand, I'm more likely to identify all the important variables in my study, making me much better protected against ambiguous results.

Putting Too Much Confidence in Convenience Data

If you deal with data on a daily basis, you've probably run across convenience data. **Convenience data** refers to a dataset that's easy to get, and it often comes from another study. For example, suppose I find a dataset from a long-term health study in which the diet and health of volunteers was tracked for several years. Some of these volunteers were on specialized diets, including a doctor prescribed regimen similar to my Type A diet. This long-term study was not designed to determine if a bar diet or shake diet is better, but because this information was tracked, I could compare the health data of the dieters to the nondieters.

In general, there's nothing wrong with convenience data. If you have it, why not use it. But there are limits to using data collected for one purpose (or no specific purpose at all) to answer a question it was never meant to answer. The person who planned the study was planning it for his or her purposes, not yours. This can cause two of the biggest *garbage in, garbage out* culprits: weakened hypotheses and confounding factors.

First, the study was not designed to answer your research questions, so it probably can't, at least not directly. You'll most likely need to weaken the original hypotheses in order to make your study fit the data-set instead of the other way around. For example, the data from the long-term health study might be thorough, and there may be a lot of it, but if the researchers didn't divide the dieters' group into a shake-drinking group and a bar-eating group, it can't be used to test my original hypothesis, that bars are better than shakes. In this case, I'd be forced to settle for a weakened hypothesis, maybe that the dieters' tend to lose more weight than the nondieters. Unfortunately, this doesn't really answer my question.

Second, convenience data has a high risk of producing confounding factors. A **confounding factor** is any variable that can confuse the outcome of a study. For example, suppose I decide weight and satisfaction with the food are the outcomes I'd like to measure when comparing the shakes to the bars. Drinking shakes or eating diet bars may well affect a person's weight, but so do exercise, diet, stress and anxiety, family history, and many other factors. Depending on how these factors were considered in the long-term health study, I may be at risk of making false conclusions. If, for example, the dieters in the long-term study were part of a group who made several lifestyle changes, including increased exercise, I have no way of concluding that it was the diet that caused weight loss and not the exercise. In other words, the exercise regimen confuses, or confounds, the results of the study.

Confusing Association and Causation

There are two basic types of studies: experimental and observational. **Experimental studies** are highly manipulated. The independent variables are carefully controlled, the dependent variables are carefully measured, and confounding factors are carefully accounted for. By controlling an experiment so tightly, researchers can establish cause and effect, prove hypotheses, and make strong scientific conclusions. Unfortunately, this type of study is not terribly common outside research and engineering laboratories.

Most studies you run across in the mainstream are **observational studies**. In an observational study, you have little or no control over variables that might affect the outcome. Online surveys are a good example of observational studies. The people collecting the data aren't trying to manipulate your opinions. They have no control over independent variables like who agrees to take the survey. All they can do is ask for your opinion and record the response.

Observational studies, even those that are perfectly planned, rarely lead to strong cause-and-effect conclusions. For example, think about an observational health study, where the eating habits of volunteers are recorded over five years. If, at the end of the study, the people who ate fish at least twice a week are healthier than those who didn't, then the researchers can claim an association between eating lots of fish and good health. That doesn't mean if *you* eat lots of fish, *you* will become healthier. It only means that, on average, the people who ate more fish tended to be healthier than their non-fish-eating counterparts.

Confounding factors are often one reason cause-and-effect conclusions can't be drawn from an observational study. Especially for human studies, where the biology is complex and the behavior even more complex, you can list every variable you can imagine and control the conditions as much as possible, but there are still uncontrollable factors that might impact the outcome of the study. For example, to establish a link between education and poverty, you can follow young people throughout their lifetimes, tracking education, employment status, and income. You can also record their race, the type of neighborhood they grew up in, the quality of their schools, and other factors that might play a role in a person's future success. But there's no way to track every variable that might impact educational or economic success, factors like a great teacher, natural talent, or an unexpected opportunity. So, there's no way to claim education causes a person to earn a higher income. At most, you can only say the two things are related.

Responsible researchers are very careful to limit the conclusions of their observational studies to associations. Unfortunately, it's all too easy to jump from association to causation, and this mistake has caused everything from strange food fads to flawed government policy over the years. Recently, two bestselling books highlighting the scandals caused by such confusion have been published. For more information and some good fun, I refer you to *Freakonomics: A Rogue Economist Explores the Hidden Side of Everything*, and *Superfreakonomics: Global Cooling, Patriotic Prostitutes, and Why Suicide Bombers Should Buy Life Insurance*, both Levitt and Dubner (2009, 2011).

REPLICATION, RANDOMIZATION, AND BLOCKING: THE BUILDING BLOCKS A GOOD STUDY

A well-designed study is built on three things: replication, randomization, and blocking. **Replication** refers to measuring multiple outcomes, not just one. For example, in a study comparing the health benefits of grape juice versus wine, multiple people would be assigned to a wine drinkers group and multiple people would be assigned to a grape juice drinkers group. The same measurements, replicates, would be taken on each person in both groups. Replication is the best way to be sure the outcomes you're observing are typical of the entire population, and not just a single, possibly unusual, individual.

Say ten of your friends are part of this wine versus grape juice study, all of them twenty-something wine aficionados, and you place them all in the wine group because you think they'll be happier there. Because they're already drinking wine regularly, these people are not be likely to show any change in health because they're simply continuing to do what they already do. Because they're all young and probably healthy, you're unlikely to observe any health problems in these people, even if drinking wine is unhealthy to the population at large. In other words, your intentions may be good, but by handpicking test subjects for the wine and grape juice groups, you just might be influencing the outcome of your study. Randomization keeps this from happening. **Randomization** is the process of assigning objects or people to groups at random, without any preconceived notions. For the wine versus grape juice study, for example, if you had fifty test subjects, you might write "wine" on twenty-five slips of paper and "grape juice" on twenty-five slips of paper, and then have the volunteers draw those pieces of paper from a hat, one at a time, in order to get their group assignments.

Blocking is the process of grouping your test subjects into subsets of homogeneous individuals. This strategy helps eliminate the effects of confounding factors so that you can focus on the impact of your independent variables. For example, suppose wine actually reduces cholesterol levels. In the wine versus grape juice study, you could assign test subjects to each group purely at random, without regard to age or gender. However, an improvement in cholesterol levels might not be as noticeable on healthy twenty-something people as it would be on overweight forty-something people, simply because the older and less healthy people have higher cholesterol levels to begin with. If you lumped all of your volunteers into two groups without regard to age and overall health and analyzed the results, the very small change in cholesterol levels of the twenty-somethings could cancel out a much larger change among the forty-somethings. And this could prevent you from concluding that wine improves cholesterol levels. In other words, age could be considered a confounding factor that confuses the outcome of this study. To eliminate the impact of this confounding factor, you could *block for age* by grouping the volunteers into decades, say 20–30, 30–40, 40–50, and 50+. For each age group, or block, you could then assign volunteers to the grape juice and wine groups at random. And then, when analyzing the results, you could compare the health outcomes within an age group, meaning comparing the twenty-something grape juice group to the twenty-something wine group, comparing the thirty-something grape juice group to the thirty-something wine group, and so on. This would take away confusion about the health benefits of wine caused by the test subject's ages. It would also give you a more detailed understanding of whether or not age plays a role in the outcome.

Replication, randomization, and blocking are the foundation of any well-planned data collection exercise. However, it's not always practical or necessary to incorporate each one into every study. When it comes to my Type A diet, for example, I have a big idea and a lot of motivation, but I also have no idea how to create a successful diet plan. Even if I limit myself to nutrition bars, I have more questions than answers. What basic recipe will give me the best flavor and texture? What vitamins do I need to add to make them nutritionally complete? What flavors should I incorporate to make them appeal to the typical Type A dieter? To answer all these questions, I could run a series of massive studies, complete with randomization, replication, and blocking: one to formulate a basic recipe from all possible combinations of ingredients, one to optimize the nutritional value of the bar, and one to decide on a small number of flavor varieties to put on the market. But with so many questions and a very limited budget, I can't possibly do this. So, I must rely on less-than-perfect data for some of my research.

I may have more enthusiasm than knowledge when it comes to developing a weight loss plan, but I also have a lot of experience in conducting research. I know there are three basic types of research: exploratory, descriptive, and experimental. I know that when you set out to answer a series of questions, you start small and flexible with exploratory research and work your way toward more formal, structured methods like those in experimental research. And I know there are different strategies for incorporating replication, randomization, and blocking into the different types of research. This will help me minimize time and cost while at the same time developing a diet program that will be a success.

Here's what I know.

EXPLORATORY RESEARCH: GETTING YOUR STUDIES INTO FOCUS

Pretty much all research starts on a computer or, if you've yet to merge onto the information superhighway, in a library. There's a lot of information out there, so no matter what you're researching, chances are somebody else has already thought about, looked into, and published something related to it before. Learning about what people have done before you can really help you focus your attention on the things that are important in your own studies, and it's an important part of your initial research.

Exploratory research, the process of learning as much as possible about your subject matter, is a crucial step in the research process. In addition to background reading, exploratory research includes data collection and ad hoc statistical analysis. The purpose at this stage isn't to answer a single, targeted question. Rather, it's to figure out what questions are relevant and identify what variables will be important when you eventually try to answer them definitively. For example, I know I want to create a line of tasty and healthy nutrition bars for my Type A diet. But that's all I know. I need the kind of direction and focus that exploratory research can give.

Exploratory research tends to be flexible and open-ended. You can find out what has and has not worked in the past. You can use it to formulate a specific hypothesis or question. You can identify dependent variables, uncover independent variables, and learn about potential confounding factors. You get to gather and analyze data in any way that makes sense, and you can even switch focus midstream. This makes exploratory research the hands down favorite of all the scientists I work with, but it also keeps you from taking what you've learned and making precise and reliable conclusions about the population you're studying.

Take my Type A diet bars. There are a lot of nutrition bars already on the market. Many companies have tried, and many have failed, to make a bar that people want to eat. To get an idea of what's available, I go to a handful of supermarkets and health food stores, buy every nutrition bar I can find, and take an inventory of their ingredients. The US government has established recommended daily amounts of all the important vitamins and minerals, and based on what I see, most health bars are loaded with about the same nutritional lineup. So it's not nutritional value that differentiates one from the next. It's got to be taste.

Already I'm able to focus my research. No need to run a study optimizing the nutritional value of my diet bars. That's pretty much already been done by the government and the food industry. I'll simply partner with a vitamin company that can supply me with a nutritionally complete powder to add to the bars and move on. This way, I can concentrate my research time and dollars on coming up with a recipe that appeals to the Type A personality.

I'm not one to let food go to waste. Especially when exploratory research is on the line. So, I gather a few friends and we taste every nutrition bar I bought. We write down our observations, carefully recording what we liked and what we didn't like about each one. Then we compare notes and come to some sort of consensus.

This food-tasting exercise is a data collection effort, and so you can call it a study. But it's extremely informal and woefully lacking in proper design. Because I've invited friends to join me, there is some replication, but there's no randomization and no blocking to prevent bias in the outcome. But at this stage, I don't really care. I'm not trying to statistically characterize the flavor and texture of the entire population of nutrition bars, I'm simply trying to get an idea of the best the market has to offer.

Once I've recovered from the gut bomb caused by tasting all those bars, I'm ready to go over my notes. There are a couple of themes that keep emerging. First, many of the bars taste like vitamins and have a gummy, artificial texture. Second, none of us thought we could stay on a diet consisting only of the nutrition bars we tested. These two observations cause me to make a declaration. No artificial, flavorless pressed food-like bricks for my weight loss plan. If they're anything like me, the health-conscious Type A personality wants a natural diet bar that looks and tastes like real food.

This helps me focus my research further. I decide to forgo the recipe-in-a-laboratory approach to formulating my basic nutrition bar. Instead I refer to the ultimate source for delicious, real, old-fashioned food: my grandmother's recipe box. There's a top secret oatmeal bar recipe in there made from whole oats and sweetened with honey, and from what I remember, it's delicious. I take this recipe and make it more nutritious by adding the vitamin powder. The result tastes just like Granny used to make, and it's a good start. But it's also a little unsophisticated for today's tastes. To make this bar appeal to the modern Type A and to add some variety, I need to spice it up with some new ingredients.

The Internet can be extremely useful when doing exploratory research, and it's a great place to gather convenience data. Websites like Amazon.com

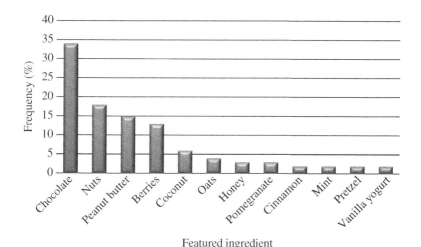

FIGURE 2.1 Popularity of ingredients in Amazon top 100 ranked nutrition bars.

allow you search for a specific item and then sort the results from most popular to least popular. They also post customer reviews and rankings. At the time of this writing, Amazon has over a thousand products fitting the search term "nutrition bars." In order to see what people are currently buying, I sort the results by customer ranking and peruse the first twenty pages or so. Figure 2.1 shows a bar chart of the most popular featured ingredients in the top 100 ranked bars. Keep in mind that most of the nutrition bars had more than one ingredient, chocolate peanut butter, for example, or cranberry macadamia nut, and so most of the bars are represented more than once in this bar chart. In other words, this isn't a frequency distribution (see Chapter 4 for more on this), it's simply a ranking of ingredients from most popular to least popular.

Here are some of my observations:

- Chocolate is king. Chocolate brownie, dark chocolate, chocolate chip, chocolate mint, chocolate raspberry, chocolate almond, and other cocoa-containing flavors dominate the market.
- Of all the tree nuts, almonds appears to be the most popular.
- Peanut butter (which is technically not a nut but a legume) is also very popular.
- Cranberry seems to be the most popular fruit addition.

Remember, this is just exploratory research. I only went to one website, and I relied on Amazon's formula for customer rankings to single out the most popular bars on the market. Also, my results only reflect the opinion of the most vocal customers, those who felt strongly enough to write a review. I have no idea how many of them are Type A personalities (though I suspect a good number of them). In any case, my data could hardly be considered a proper random sample representing the typical nutrition bar customer. Despite its flaws, this convenience dataset gives me a general picture of the market, and it's a great place to start.

I don't want my diet bars to be just another drop in an ocean of chocolate peanut butter, honey oat, and cranberry almond. So I decide to stay away from the ingredients in Figure 2.1. Except for chocolate, that is. In my mind, you can never have too much chocolate. As for the other ingredients, if I want my bars to stand out, dieter they need something different. Something highly nutritious and tasty. Something that will appeal to the supermen and superwomen of today.

How about superfoods?

Superfoods, foods that deliver a lot of nutritional bang for the buck, are said to boost your energy and your health. Fish, green vegetables, nuts, berries, and seeds are examples of superfoods. These are the perfect ingredients to launch my nutrition bars to the top of the heap. Unfortunately, I'm not enough of a chef to turn foods like salmon and broccoli into a tasty and appealing granola bar. So certain items immediately drop off my list. Blueberries, oats, and nuts are already a mainstay of the nutrition bar market. These will make good basic ingredients, but won't make my product stand out from the rest of the pack. Instead I turn to the many underutilized superfoods, things like chili peppers, chia seeds, and acai berries. If I incorporate them with some of the more mainstream ingredients, for example, yogurt, and nuts, and chocolate, I have a list of possible ingredients to combine into what I'm now calling the ultimate Type A superfoods bar.

DESCRIPTIVE AND EXPLANATORY RESEARCH: ANSWERING THE TARGETED QUESTIONS

The exploratory research I did was well worth the time it took. I was able to eliminate a couple studies from my list and focus my attention on an outcome: flavor. My initial taste testing gave me a good idea what not to do. And the convenience data I gathered helped me narrow down my independent variables. It's time to finish off my recipes and see what potential customers think. I can do this with the help of descriptive and explanatory research.

Unlike exploratory research, **explanatory research** is structured and targeted to a specific goal. Explanatory research typically involves controlled

	Exploratory	Descriptive	Explanatory
Types of studies	Ad hoc, anything that makes sense	Observational	Controlled experiments (factorial, controlled trials, repeated measures)
Strength of conclusions	Weak, not generalizable to an entire population	Usually limited to associations	Associations and/or cause and effect

FIGURE 2.2   The three types of research.

experiments, those studies that allow you to manipulate independent variables and draw cause-and-effect conclusions from them. A controlled experiment where I manipulate the ingredients in my nutrition bars and observe the impact on flavor would be considered explanatory research. **Descriptive research** sits in the middle, somewhere between exploratory and explanatory research. The purpose of descriptive research is merely to describe, not to establish cause and effect, and it typically involves observational studies. A market survey like the one summarized in Figure 2.1 could be considered descriptive research. I ran an observational study on the featured flavors in popular nutrition bars. I didn't try to establish why chocolate is the king of all nutrition bar ingredients, I only observed it.

Figure 2.2 lists the three different types of research and the studies that go along with them. These studies are described in detail throughout the remainder of this section.

Controlled Experiments: The Art of Manipulation

There's something very satisfying about running what's called a controlled experiment. In a **controlled experiment**, you get to manipulate the independent variables and then observe the outcome. For example, suppose I wanted to minimize the amount of sugar in my nutrition bars without sacrificing taste. I might make several versions of the nutrition bar, each with different amounts of sugar, and have a panel of taste testers give me feedback on the result. In this experiment, the amount of sugar is what we call a **controllable variable**. It's an independent variable that's systematically manipulated to allow me to observe its impact on the outcome. It's direct. It's conclusive. It's satisfying.

Some experimental studies are as straightforward as the sugar study I just mentioned. But in the real world, most are not. Often you'll have many independent variables, some of them controllable and others uncontrollable.

An **uncontrollable variable** is just what the name implies, an independent variable that cannot be systematically manipulated. The heat of a particular batch of chili peppers, whether or not a randomly chosen taste tester prefers subtle to strong flavors, these are uncontrollable variables. And these can make controlled experiments a real challenge to plan and run. Fortunately, there's a branch of statistics, **design of experiments (DOEs)**, that lays out strategies for setting up different types of controlled experiments. Every study is unique, with different dependent and independent variables and different potential confounding factors, but most experiments fall into one of a small number of categories. Three of the most common are mentioned here.

Factorial Studies   I have my basic nutrition bar recipe, and I have a list of ingredients I'd like to throw in. How should I combine these ingredients to give me the tastiest possible flavor combinations? In planning a study to answer this question, I'll use a factorial design. **Factorial designs** are often used for experiments with a single dependent variable and lots of independent variables, each with a small number of possible values, or **levels**. By giving you a way to manipulate the independent variables in a systematic and scientific manner, factorial designs can simplify the process of planning an experiment with many independent variables.

For example, say I've settled on dark chocolate, jalapenos, chia seeds, acai berries, and almonds as potential add-on ingredients to my nutrition bar. I could combine these ingredients in different ways, and I'd like to run a study to determine which combinations give me the best flavor. In this study, there's a single dependent variable, taste. There are five independent variables: dark chocolate, jalapeno peppers, chia seeds, acai berries, and almonds. Each independent variable, or ingredient, can be added in any amount, from none all the way up to tastebud overload. To start with, I'll restrict the levels, or amounts, of each ingredient to two: zero for none added, one for a nominal amount added.

Five possible ingredients. Two possible amounts for each. If you start listing all the combinations, it won't take long to realize there are many. To calculate exactly how many, you take the number of levels (amounts added) and raise it to the power of the number of independent variables (ingredients). With five ingredients and two amounts for each, there are a total of $2^5 = 32$ possible combinations. That may be enough to scare away a typical laid back, Type B personality, but not an overachiever like me. I'll test them all using a **full factorial design**. This design is listed in Figure 2.3.

The great thing about a full factorial design is that it allows you to not only test the impact of each independent variable but also to look for what are called **interactions**, the impact of combinations of independent variables. For

No add-ins	Chocolate, almond	Chocolate, jalapeno, almond	Chocolate, jalapeno, chia, almond
Chocolate	Jalapeno, chia	Chocolate, chia, acai	Chocolate, jalapeno, acai, almond
Jalapeno	Jalapeno, acai	Chocolate, chia, almond	Chocolate, chia, acai, almond
Chia	Jalapeno, almond	Chocolate, acai, almond	Jalapeno, chia, acai, almond
Acai	Chia, acai	Jalapeno, chia, acai	Chocolate, jalapeno, chi, acai, almond
Almond	Chia, almond	Jalapeno, chia, almond	
Chocolate, jalapeno	Acai, almond	Jalapeno, acai, almond	
Chocolate, chia	Chocolate, jalapeno, chia	Chia, acai, almond	
Chocolate, acai	Chocolate, jalapeno, acai	Chocolate, jalapeno, chia, acai	

FIGURE 2.3   Full factorial design for type A superfoods bar study.

example, chia seeds alone may give my nutrition bars a birdseed like quality. But adding acai berries may counterbalance the seeds, giving my bars a pleasant, sweet and crunchy combination. In other words, the chia seeds and acai berries may interact to give me a much better outcome than either of the two ingredients alone.

On the down side, when you have many independent variables or more than two levels, full factorial designs can rapidly become unwieldy. With five ingredients and two possible amounts, there are thirty-two possible combinations to test. If I add just one more level, or amount, I now have five independent variables and three levels for a total of $3^5 = 243$ combinations. If I now add just one more ingredient to the list, the number of combinations becomes $3^6 = 729$, before even thinking about replication. That's enough to scare even the most A-like of the Type A researchers out there.

It's important to get a complete dataset out of a study, but it's also important to recognize when such a dataset is impractical. Over the years, different strategies have been developed for reducing full factorial designs down to a more reasonable size. Limiting the number of levels to two, for example. This strategy works well for screening experiments, experiments that help you narrow down the list of combinations to those having the most impact. Screening experiments are typically followed by another study focusing on a much smaller list of independent variable combinations. For my Type A superfoods study, for example, the two-level full factorial design could be considered a screening experiment that allows me to identify the tastiest add-in combinations. Most likely, I'd run a follow-on study in which I play with the amounts in the winning combinations in order to get the best flavor.

In addition to screening experiments, there are variations of the full factorial design that can greatly reduce the number of combinations to be tested. One of the most common variations is the fractional factorial design. A **fractional factorial** design takes a carefully chosen subset of combinations from the full factorial design and tests those. This type of experiment can greatly reduce the number of combinations to be tested, but the convenience comes at a price. With a fractional factorial design, you generally lose the ability to study something called higher order interactions. **Higher order interactions** refer to the impact of many—usually more than two—independent variables. For example, the full factorial design for my superfoods bar includes a bar made of every possible combination of the five ingredients, so I could directly observe every possible flavor combination. In a fractional factorial design, I'd lose some of the more complex flavor combinations, jalapeno, chia, and acai, for example, but because my experiment would now much smaller and cheaper to run, it's probably worth it.

The basic idea behind fractional factorial designs is this. Suppose I can't test all $2^5 = 32$ flavor combinations in my original full factorial design. Let's say I can only afford to test half of them. A fractional factorial design carefully selects sixteen combinations so that (i) all main effects (the impact of the individual variables) can be measured, (ii) all two-way interactions (the impact of two-variable combinations) can be measured, and (iii) the design is balanced (the number of tests having each level of each variable is the same). Designing fractional factorial experiments is a detailed process, and well beyond the scope of this book. Fortunately, there are many resources out there on the subject, and there are tables that do most of the work for you. If you find yourself in need of such a design, I refer you to *The Design and Analysis of Experiments* by Montgomery (2012).

Once I have the flavor combinations to try, planning the rest of the superfoods bar taste test is fairly straightforward. I need some replication, so I'm not fooled by a single, possibly eccentric taste tester. I form a panel of five people, each a professional taste tester with a discriminating palette. Each person will try every bar and rate it. But how? The order in which each person tastes the bars matters. For example, if the chocolate jalapeno bar always comes before the much milder acai almond bar, the heat of jalapeno peppers might spoil the taste of the bar that follows it. There are two ways I could handle this: randomization and blocking. Randomization would include randomizing the order in which each taste tester tastes the bars, making sure no two bars follow one another for more than one individual. That way, if strong flavors spoil the flavor of a bar for one taste tester, I have four more opinions to counteract it. Blocking might involve dividing the bars into blocks, where each block has similar flavor profiles ordered from mildest to strongest, and having the taste testers try the bars one randomly chosen block at a time. In this way, the taste testers get a break, and I can be confident the bars are ordered so that the impact will be minimized. Because I only have five taste testers and I need objective opinions from each and every one, I decide to go with blocking.

There's one final but important consideration in designing this experiment, and it's familiar to every data analyst who's ever had to analyze a poorly designed survey. I could give the taste testers blank paper and have them write down their impressions of each bar as they go. But if I give five different taste testers the freedom to respond in any way they'd like, I'll probably get five very different responses. For example, the responses to the jalapeno chocolate bar could quite possibly look like the following:

The chili pepper is too strong.

Jalapeno and chocolate are a great combination.

Needs more chocolate.

I'd prefer milk chocolate to dark chocolate.

Loved it!

Don't get me wrong. These comments are constructive and can be a real help when tweaking the amounts of each ingredient. But there's no way to combine these responses in a way that can be statistically (and numerically) analyzed.

When humans are used to measure outcome, it's important to focus their feedback, and it's helpful to have them give their responses in a way that's objective and easy to analyze. A scale of one to five, called a **five-point Likert scale**, is a very common and effective way to do this. To use this Likert scale, I ask a set of questions and provide five multiple choice responses for each. My questions look something like those shown in Figure 2.4. This gives me a more objective way to evaluate the responses, and helps me get feedback on specific areas of concern. For more information about designing a good survey, I refer you to *Designing and Conducting Survey Research: A Comprehensive Guide* by Rea and Parker (2005).

	Strongly disagree	Disagree	Neutral	Agree	Strongly agree
1. The texture is pleasant	1	2	3	4	5
2. The flavors work well together	1	2	3	4	5
3. The flavors are well balanced	1	2	3	4	5

FIGURE 2.4   Type A superfoods bar survey questions.

Controlled Trials   Controlled trials appear in laboratory animal studies, drug studies, and weight loss studies, just to name a few examples. In a controlled trial, you'd like to test the impact of some independent variable, called a **treatment**, on your outcome. You do this by dividing your test subjects or items into two groups. One group is subjected to the treatment, and the other isn't. Or, if you're interested in different levels of treatment, like different doses of a drug, you select as many groups as treatment levels and assign a different treatment level to each one. At the end of the experiment, you compare the groups to one another and make conclusions about the impact of the treatment.

Say I've completed the superfoods nutrition bar study and settled on three flavors: chocolate acai, chocolate jalapeno, and chia almond. The bars are ready. Now it's time to see if my diet works. In this case, I might run a controlled trial where I select some number, say 100, volunteers, and split them into two groups. One group follows my diet and the other follows a popular competitor's diet. At the end of some period of time, say two weeks, I compare the **control group**, those individuals who followed the other guy's diet, to the **treatment group**, those people who followed my diet. If the treatment group has lost more weight than the control group, I can declare success.

Controlled trials are fairly easy to understand, but they're not always easy to design. There are often confounding factors to be taken into account, especially when you're experimenting on humans. When comparing my diet to the competition, for example, I need to recognize that age, activity level, pre-existing health conditions, and gender all play a role in a person's ability to lose weight. And if I choose my control and treatment groups poorly, say, by placing all the twenty-something fitness nuts in the treatment group, the results will most likely favor my diet, not because it's better, but because of how I chose my sample.

Replication, randomization, and blocking are particularly important for controlled trials. Proper replication gives you something called statistical power (see Chapter 3), which increases the likelihood you'll be able to measure real differences between the treatment and control groups. Assigning test subjects to groups at random reduces the risk you'll bias your results by, for example, placing all the young weekend warriors in the same group. Blocking removes confounding and reduces variation within each group, helping you to see differences more clearly. For my weight loss study, I might divide test subjects into blocks based on three fitness and activity levels: none, moderate, and high. Within each block, I'd then assign them to control and treatment groups at random. This would give me more homogeneous groups with potentially smaller variation in the outcome, and it would allow me to study the impact of exercise on my diet program.

Repeated Measures (Before and After Studies) Suppose I'd like to know the overall effectiveness of my Type A diet, without regard to my competitors. In this case, I could gather some volunteers, weigh them, and instruct them to follow my diet for two weeks. At the end of the study, I could weigh the volunteers again to see how much each one lost. This type of before-and-after study is called a **repeated measures** study because the test subjects are measured several times over the course of the trials. Repeated measures studies are different from controlled trials because you are not comparing one group of test subjects (the treatment group) to another (the control group). Instead, you're comparing each test subject to himself or herself, looking for changes caused by the treatment.

Repeated measures studies can be very enlightening. By comparing before-and-after data on a group of test subjects, you can minimize confounding factors. In other words, you can directly observe the impact of a treatment without other variables confusing the outcome. But problems can still arise. Humans are the ultimate uncontrollable variable, and uncontrollable variables can wreak havoc on any study. So it's important to observe and record anything that might impact the outcome. In a weight loss study, for example, things like major life changes, illness, exercise, and other factors could impact a person's ability to follow the diet and lose weight, and so it's important to record them all.

Observational Studies: Scientifically Approved Voyeurism

By manipulating independent variables in a controlled experiment, researchers can establish cause and effect, prove hypotheses, and make strong scientific conclusions. Unfortunately, many studies you run across are observational studies. In an observational study, you have little or no control over independent variables that might affect the outcome. Market research surveys are observational studies. Researchers have very little control over who agrees to participate in the survey, and they try not to manipulate your opinions (at least, not if they want an honest assessment). All they can do is offer free stuff to entice you to take the survey, ask nonleading questions, and observe the result.

The most important consideration in setting up an observational study is choosing your sample. **Sampling** is the science of choosing a subset, or sample, for a study. Manufacturing engineers use sampling techniques to pick which parts coming off an assembly line should be tested for defects. Medical researchers use it to study the prevalence of diseases and the effectiveness of drugs. Market researchers use it to gather customer likes and dislikes. Survey sampling is a broad area of statistics, and there are entire classes devoted to the subject. I get a single section to talk about it, so I'll only review a few of the most common sampling techniques. For more information, I refer you to *Sampling* by Thompson (2012).

Simple Random Sampling   Simple random sampling is the most straightforward way to capture a sample for your study. With this sampling scheme, you choose the size of your sample and then select your test subjects (or parts or items) completely at random. Every member of the population has an equal probability of being selected. This sampling scheme doesn't guarantee the outcome of your study will be perfectly accurate. But by choosing samples at random, you minimize the chances of bias due to more subjective approaches. For example, suppose I want to repeat the market analysis I did in my exploratory research, where I ranked nutrition bar ingredients by popularity, only this time I want to perform a comprehensive analysis of the entire market, not only those bars most highly ranked by Amazon.com customers. In this case, I would list all bars offered by all my competitors and pick a random subset of those for taste testing and follow-on analysis. This process of random sampling would help me understand the entire market, not just the preferences of Amazon customers.

Systematic Sampling   Sometimes, simple random sampling just doesn't make sense. For example, suppose I take my diet bars to the streets in order to get feedback on their taste. I could take a map of the entire metropolitan area, choose houses at random, and elicit feedback from those houses. Driving around the city, knocking on doors, and trying to entice people to try my Type A diet bars would be expensive and time-consuming. Having so many doors slammed in my face would be painful and humiliating. Definitely not worth the effort.

Systematic sampling is an alternative to simple random sampling when it makes sense to collect data in a more orderly fashion. For example, rather than driving all over the city and begging people to try my nutrition bar, I might pick a four block radius and select every other house. By doing this,

I spend less time driving and more time getting feedback from potential customers. Systematic sampling has its limitations, however. If, for my door-to-door study, I happen to choose a strongly ethnic neighborhood, one whose cuisine tends toward spicy foods, the feedback on my mildly sweet nutrition bars could be more negative than it would be had I chosen a different neighborhood.

Quota Sampling   Suppose I have reason to believe men and women will respond differently to the Type A diet. In this case, it makes sense to treat them separately in my study. Rather than taking a purely random sample of test subjects, I'd set quotas, a specified number of men and women, and gather volunteers until I had enough of each. This is called **quota sampling**. Quota sampling is often used when you want to make sure some subgroup is adequately represented in your study, especially when that subgroup comprises a very small percentage of the entire population. Quota sampling allows you to get enough samples in each group so you can make statistical comparisons between them. If you use this technique, however, it's important to remember you are no longer dealing with a sample that represents the natural composition of your population, and so care must be taken when generalizing results from such a study.

SO MANY STRATEGIES, SO LITTLE TIME

I've only scratched the surface of the many different strategies for planning and designing studies. There are entire courses devoted to topics covered by just a few paragraphs in this chapter, courses with names like *Research Methods*, *Design of Experiments*, and *Sample Statistics*. And there are many different ways to combine the techniques presented here when planning a study. No matter what techniques you choose or what type of study you're planning, there are three things all your well-designed studies should have: (i) a measurable set of dependent variables that address your specific question, (ii) a strategy, such as randomization and blocking, for eliminating confounding factors, and (iii) an understanding of the strength of the conclusions you can draw from your study, whether it's association or causation.

BIBLIOGRAPHY

Gilani N. Types of Designs in Research. Available at http://www.ehow.com/info_7760064_types-designs-research.html. Accessed March 24, 2014.

Harland B. Observational Study vs. Experiments. Available at http://www.ehow.com/info_8611337_observational-study-vs-experiments.html. Accessed March 24, 2014.

Lagorio C. Diet Plan Success Tough to Measure. http://www.cbsnews.com/news/diet-plan-success-tough-to-weigh/. Accessed January 3, 2005.

Levitt SD, Dubner SJ. *Freakonomics: A Rogue Economist Explores the Hidden Side of Everything*. New York: Harper Collins; 2009.

Levitt SD, Dubner SJ. *Superfreakonomics: Global Cooling, Patriotic Prostitutes, and Why Suicide Bombers Should Buy Life Insurance*. New York: Harper Collins; 2011.

Montgomery DC. *Design and Analysis of Experiments*. New York: John Wiley & Sons, Inc.; 2012.

Rea LM, Parker RA. *Designing and Conducting Survey Research: A Comprehensive Guide*. San Francisco: John Wiley & Sons, Inc; 2005.

Rossiter C. Differences Between Descriptive and Exploratory Research. Available at http://www.ehow.com/info_8360417_differences-between-descriptive-exploratory-research.html. Accessed March 24, 2014.

Thompson SK. *Sampling*. Hoboken: John Wiley & Sons, Inc.; 2012.

United States Department of Agriculture Dietary Reference Intake Tables, Available at http://fnic.nal.usda.gov/dietary-guidance/dietary-reference-intakes/dri-tables. Accessed August 11, 2014.

Zelman K. 10 Everyday Superfoods. Available at http://www.webmd.com/food-recipes/features/10-everyday-super-foods. Accessed April 23, 2014.

3

CONSERVATIVES, LIBERALS, AND OTHER POLITICAL PAWNS: HOW TO GAIN POWER AND INFLUENCE WITH SAMPLE SIZE CALCULATIONS

The 2012 US presidential election was full of drama. Conservatives beat up on incumbent Democrat Barack Obama, claiming a weak economy and shaky foreign policy were solid reasons to boot the president out of office. Liberals hammered the Republican challenger Mitt Romney, accusing him of being too rich and cold-hearted to understand the concerns of the average American. The polls were up and down, often favoring the president but also showing steadily increasing support for the challenger as the election drew near. This political drama culminated on election night, when conservative commentator Karl Rove, reporting for FOX News, refused to believe swing state Ohio had actually gone for the president. The incredulous Rove stood up and walked to a back room behind the set to consult with the data crunchers. He only relented when they insisted they were "99.9% confident" of their predictions.

Karl Rove wasn't the only person who was confused. The political polls were confused, too. In the days leading up to the election, two of the nine major polling agencies had Romney winning, three had Obama winning, and four had the candidates in a dead heat. In the end, however, Obama won by almost 4% points, winning 51.1% of the vote to Romney's 47.2%. Not a

Beyond Basic Statistics: Tips, Tricks, and Techniques Every Data Analyst Should Know,
First Edition. Kristin H. Jarman.
© 2015 John Wiley & Sons, Inc. Published 2015 by John Wiley & Sons, Inc.

huge margin, but in this era of divided politics, it was a decided victory for the president.

Unlike the political pundits and commentators on cable news, most of us will never be in a position to embarrass ourselves on national television. For my part in that, I'm thankful. But that doesn't mean we can be careless about our data-based conclusions. The reason we collect and analyze data is because we can use it to make predictions, plan a course of action, spend money, and otherwise commit valuable time and resources. Data may be cheap, but the cost of inaccurate conclusions can be very expensive.

A well-planned study gives you the best chance at reaching accurate conclusions, so you can avoid your own Karl Rove moment. But there are no guarantees in life. All of the top nine polling organizations use carefully designed sampling strategies to capture an accurate snapshot of American voters. And yet, in the days leading up to the 2012 election, the only point they agreed upon was that nobody really knew what the outcome would be. Why? Because of uncertainty. Any time you use a sample to represent an entire population, there's a certain probability you'll end up fooling yourself. And while you can't eliminate this probability of making an error, commonly used statistical techniques can help you measure it.

These same statistical techniques can also be used to make sample-size calculations. Meant to be used as part of the study planning process, sample-size calculations help you specify the number of samples, or replicates, needed to achieve an acceptable error probability. These calculations rely on something called statistical power. In this chapter, sample-size calculations will be used to show how statistical power can be turned into political power.

STEP 1. KEEP YOUR FINGER ON THE PULSE OF THE POPULACE

Let's say I'm the senior advisor and political consultant for Steve McMann, son of a major Midwestern car dealership mogul. Steve has just broken with family tradition, deciding to trade a lifetime of comfort selling reasonably priced imports for a life of public service. In other words, politics.

Most politicians get their start by winning local elections, but my motto is go big or go home, and I convince Steve to run for the US Senate. The name McMann gives my guy a decided advantage in the race. After all, he's the heir to the McMann dealership dynasty, and his face has already appeared on commercials across the state. His opponent is some obscure city councilman from the fifteenth district, a rural part of the state that many voters have trouble pinpointing on a map. Just how much of an advantage does Steve have? A poll can tell us. We pick a large round sample size, say $N = 100$, and send a group of interns out to ask the voters of our fair state what they think of Steve. Here are the results.

Of the 100 people polled, most of the likely voters recognized the name Steve McMann, and 53% said they were inclined to vote for him. Any more than 50% is enough of the vote needed to win the election, and so 53% is good. But what if these numbers are off by a little bit? An error of just 3% and we could be looking at a major disappointment. To find the margin of error (MOE) on this estimate, I'll use one of the most common tools from basic statistics: the confidence interval.

Recall a confidence interval is a range of values inside which you can feel confident your estimate falls. Sometimes confidence intervals are reported as a range. Sometimes they're reported as the estimate plus or minus some MOE. For example, with an MOE of 3%, the number of people who favor Steve in the election might be reported as (50%, 56%) or as $53 \pm 3\%$.

There are two approaches to constructing a confidence interval for a proportion such as this one. One applies to small populations, and the other applies to large populations. When the underlying population is small, a business of 200 people, for example, the sample size will probably represent a significant fraction of the total population. In this case, special techniques based on a concept called **sampling without replacement** should be used to construct the confidence interval. When the underlying population is large, the Unites States or a large Midwestern state, for example, the population is so much larger than any sample you plan on drawing, you can assume it to be infinite. A good rule of thumb is this. When the sample size is more than a tenth of the total population size, use sampling without replacement techniques to estimate your uncertainty (see Barnett, 2002). Otherwise, use a method like the normal approximation described here.

For large sample sizes ($N > 25$) drawn from a very large population, and for an estimated proportion, call it $\hat{p}$, where $N\hat{p} > 5$ and $N(1 - \hat{p}) > 5$, the confidence interval for $\hat{p}$ can be calculated with the following equation:

$$\hat{p} \pm z_{1-\frac{\alpha}{2}} \sqrt{\frac{\hat{p}(1 - \hat{p})}{N}}$$

where $z_{1-(\alpha/2)}$ is the $1-\alpha/2$ critical value for the standard normal distribution (see Appendix A). In my case, $\hat{p} = 0.53$, and $N = 100$, so both $N\hat{p}$ and $N(1 - \hat{p})$ are greater than five. Appendix A gives a value $z_{0.975} = 1.96$ for a 95% confidence interval, so the estimated voter approval rating is

$$0.53 \pm 1.96 \sqrt{\frac{0.53 \times 0.48}{100}} = 0.53 \pm 0.10.$$

This means Steve's approval rating lies somewhere inside the range 43–63%. In other words, after all the work of writing the poll, carefully selecting the best polling locations, sending out the interns, and crunching the numbers, I still have no idea if my guy is ahead of the crucial 50% mark in the upcoming election.

STEP 2. AVOID AMBIGUOUS RESULTS AND OTHER POLITICAL POTHOLES

Most data analysts who've been out in the real world very long have run into this situation. A carefully designed study, faithfully executed with a nice random sample, and nothing but ambiguous results to show for it. When this happens, sample size might be the problem. The amount of uncertainty you have depends on the sample size, specifically, the bigger the sample, the smaller the uncertainty. To avoid falling into a pothole of statistical ambiguity, you need to have enough samples so that the uncertainty will be small enough to yield practically meaningful conclusions. Sample-size calculations help you determine a sufficient sample size *before* embarking on a time-consuming and expensive study.

Most researchers who come to me asking about sample sizes expect me to produce a number off the top of my head without much knowledge of what they're planning to do. Unfortunately, these people are always disappointed. Calculating sample sizes isn't as easy as punching a few numbers into your computer. Sure, there are statistical techniques designed to calculate samples sizes, there are even free online calculators for this purpose. But before you use them, you need a solid understanding of the goals and limitations of the study you're planning. Here are the important steps in the sample-size calculation process.

Identify the Data Analysis Technique You Will Be Performing

It may seem strange to think about analyzing the data before it's been collected, but this is the time to do it. You're collecting data not just for the sake of collecting data, but to analyze it in a way that answers your question. So it makes sense to think about the techniques you'll use before you start your study. Will you be running a hypothesis test? Constructing a confidence interval? Performing regression? Answering these questions can help you focus your study. And as you'll see in the next section, it's also a necessary step in identifying the right tools to use to calculate samples sizes.

Know the Difference Between Practical Significance and Statistical Significance

Most people with a basic statistics course under their belt are at least somewhat familiar with the concept of **statistical significance**. When performing a hypothesis test, for example, if the p-value is below the critical error probability you specify, say 0.05, then the results are statistically significant. In other words, the data are inconsistent with the null hypothesis to the extent that statistically, you'd reject H_0 in favor of the alternative.

Fewer data analysts are familiar with the concept of practical significance. **Practical significance** is the result that matters to your particular problem. For example, for a NASA engineer looking to get his astronauts safely to Mars, an MOE of three units might be way too big to guarantee the spaceship holds together throughout the journey. To a politician looking for a ballpark figure on approval ratings, a margin of 3% points may be plenty.

Here's another way to think of practical significance. Suppose you're running a hypothesis test for the mean of a population. Your hypotheses might be:

$$H_0 : \mu = 0 \text{ vs.}$$
$$H_A : \mu > 0.$$

This test determines whether the mean of your population is greater than zero or not. But because of the specifics of your particular problem, you may not care if the mean is small, say $\mu = 0.5$. A mean value this small may have no practical significance to you. It may only matter if the mean is, for example, $\mu = 1$ or greater. In this case, the practically significant mean isn't what's laid out in the hypothesis test, it's something determined in the context of your study.

Statistical significance and practical significance can be two very different things. Suppose the true mean of the population is a practically insignificant value of $\mu = 0.5$. Depending on the outcome of your study and the number of samples you collect, it's entirely possible your data analysis will come back with a rejection of the null hypothesis. In this case, the results are statistically significant but not practically significant. On the other hand, suppose the true mean of your population is $\mu = 1.5$. This is a practically significant value. But if you have a lot of variation in your data and a small sample size, the hypothesis test may very well come back with a p-value that's not statistically significant.

Note Your Practical Limitations

In my experience, practical limitations are typically just as important to sample-size calculations as statistical considerations. Studies take time and cost money, and it's not always possible to collect as many samples as you (or your sample size calculator) would like. Therefore, before planning a study, it makes sense to have a clear idea of how many samples is too many, practically speaking. This can help guide the experimental design from the beginning, keeping you from wasting your time planning an extensive study that could never be run.

STEP 3. LET SAMPLE-SIZE CALCULATIONS BE YOUR RIGHT-HAND MAN

With Steve McMann's senatorial campaign in motion, it's time to boost our efforts. This means an image consultant, campaign ads, speeches, and unplanned stops at strategic locations around the state to show the voters he's one of the people. It also means better polling. We need to know where Steve stands with the voters, and if we want to track his poll numbers with any degree of accuracy, we need a larger sample size. Just how much larger can be determined using sample-size calculations.

Population Means and Probabilities: Sample-Size Calculations for a Confidence Interval

Whatever statistical technique you plan on using, the approach to calculating the sample size is the same: (i) take a guess at some necessary estimates, and (ii) plug those estimates into the appropriate formula and solve for the sample size N. With regards to confidence intervals, the estimates might include the sample mean and variance or a proportion, a confidence level, and an acceptable MOE. The appropriate formula is the formula for the confidence interval you plan to use. With enough statistical sophistication, you can use this approach for a confidence interval around any estimate—median, standard deviation, range, and so on. However, since confidence intervals for a mean and a proportion are by far the most common, I'll restrict my attention to those. The details on how to construct these two types of confidence intervals can be found in most basic statistics textbooks. In this chapter, I'll simply show how they can be used to determine the sample size for a study.

Suppose I'm ready to launch a second poll of likely voters across the state, this time with a much larger sample size. Do the voters favor Steve McMann over the no-name city councilman from the fifteenth district? As before, my main objective in this study is a confidence interval on the percentage of likely voters that favor Steve over his opponent. The first study was woefully under-sampled, but I can still use those results to help me determine N. Specifically, suppose the estimate in that study was reasonable and Steve's approval rating is about 53%. I'd like to know if this value is significantly larger than 50%, in other words, if it has an MOE of 3% or less. This is my practical significance.

The confidence interval for a single proportion is

$$\hat{p} \pm z_{\frac{\alpha}{2}} \sqrt{\frac{\hat{p}(1-\hat{p})}{N}},$$

so the MOE on this confidence interval is

$$\text{MOE} = z_{\frac{\alpha}{2}} \sqrt{\frac{\hat{p}(1-\hat{p})}{N}}.$$

Solving this equation for N gives me

$$N = \hat{p}(1-\hat{p}) \left[\frac{z_{\frac{\alpha}{2}}}{\text{MOE}} \right]^2.$$

With a favorability rating of 53% in the polls, $\hat{p} = 0.53$. To achieve an MOE of 3%, MOE$=0.03$. And for a standard two-sided 95% confidence interval, $z_{(\alpha/2)} = z_{0.025} = 1.96$. This gives me everything I need to calculate N,

$$N = 0.53 \times 0.47 \times \left[\frac{1.96}{0.03} \right]^2 = 1064 \text{ samples.}$$

My first poll had only a hundred likely voters in it, but this calculation suggests I need more than ten times that number to achieve the practically meaningful MOE of 3%.

This sample size formula works for any study designed to estimate a proportion. However, in some cases, you might have no idea what to plug in for $\hat{p}$. After all, estimating this value is the whole point of the study. Rather than taking a wild guess, you can use a shortcut. In particular, the first term in the sample size formula—the quantity $\hat{p}(1-\hat{p})$—directly impacts the sample size calculation. The larger the quantity $\hat{p}(1-\hat{p})$, the larger the number of samples required to achieve a given MOE. If you plot $\hat{p}(1-\hat{p})$ as a function of the value $\hat{p}$, you obtain Figure 3.1. This value starts at zero when $\hat{p} = 0$, gradually rises to its maximum point at $\hat{p} = 0.5$, and then gradually

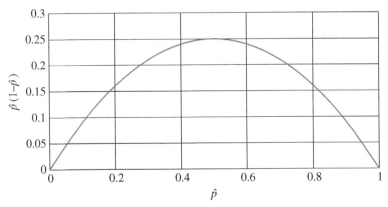

FIGURE 3.1   How the confidence interval for a proportion depends on $\hat{p}$.

falls back to zero at $\hat{p} = 1$. Rather than taking a wild guess at a likely value of $\hat{p}$, you can take a conservative approach and plug in the maximum possible value of $\hat{p}(1-\hat{p})$, which is 0.25. This shortcut gives you the maximum number of samples required to achieve your specified MOE, in other words, a sample size that guarantees you'll get a confidence interval to your liking. Here's the approximation to use when taking this conservative approach:

$$N \approx \frac{1}{4}\left[\frac{z_{\frac{\alpha}{2}}}{\text{MOE}}\right]^2.$$

Plugging an MOE of 0.03 and $z_{\alpha/2} = 1.96$ into this equation gives $N = 1065$. Not surprisingly, this is very close to the value of $N = 1064$ calculated from the proportion $\hat{p} = 0.53$.

This same approach can be used to calculate the sample size for any type of confidence interval. For example, the formula for a confidence interval around the sample mean, with unknown variance, is

$$\bar{x} \pm t_{\frac{\alpha}{2}, N-1}\frac{s}{\sqrt{N}}$$

Where $\bar{x}$ and s are the sample mean and sample standard deviation, $t_{1-(\alpha/2), N-1}$ is the $1 - (\alpha/2)$ critical value for the Student t-distribution with $N-1$ degrees of freedom (see Appendix B), and N is the sample size. The MOE for this confidence interval is

$$\text{MOE} = \pm t_{\frac{\alpha}{2}, N-1}\frac{s}{\sqrt{N}}$$

Solving for the sample size N gives

$$N = \left[t_{\frac{\alpha}{2}, N-1} \frac{s}{\text{MOE}} \right]^2$$

Note here, the critical value $t_{(\alpha/2), N-1}$ changes with the sample size N, and this complicates the formula. However, for larger studies, when the sample size is $N > 20$, the t-value doesn't change much. For example, as shown in Appendix B, the $\alpha = 0.05$ critical t-value hardly changes between $N = 20$ and $N = 100$ observations. So, for purposes of calculating sample sizes, an approximate t-value, such as $t_{0.025, N>20} \approx 2.0$, can often be used.

By making this approximation, taking a guess at the sample standard deviation s, and putting in the desired MOE, an approximate sample size can be determined. For example, suppose your initial investigation showed that the standard deviation of your sample is about 3.0. For an MOE a third of that value, or 1.0, and a 95% confidence interval, you'd plug $s = 3.0$, MOE $= 1.0$, and $t_{(\alpha/2), N-1} \approx 2.0$ into this equation to get $N \approx 36$ samples needed.

Power and Sample-Size Calculations for Hypothesis Tests

When your data analysis plan calls for a hypothesis test, the idea behind determining sample size is similar to that for a confidence interval: decide what test you'll be using, take a guess at any estimates you need for the test, determine a practical significance level you can live with, and calculate the sample size. The idea is similar. But the process is different.

Recall that every hypothesis test has two parts: a test statistic and a decision criterion. The **test statistic** is a value calculated from the data. This value carries evidence for or against H_0, in other words, it's what you use to judge whether or not the null hypothesis is true. The **decision criterion** is a threshold. You compare the test statistic to this threshold in order to decide whether you should accept that H_0 is true, or reject H_0 in favor of H_A. For a more complete review of basic hypothesis testing, I refer you to a good basic statistics text.

When running a hypothesis test, there are two types of errors you can make. First, you can reject the null hypothesis when it's actually true. This is called the **type I error**. Second, you can accept the null hypothesis when it's actually false. This is called the **type II error**. For any hypothesis test, there's a chance you'll make a type I or type II error. The probability of making a type I error is a value you specify when setting up the test. It's the significance level, the alpha value, $\alpha = 0.05$, for example. The type II error probability, usually referred to by the parameter β, is not something you specify when you set up a typical test. It simply comes along for the ride.

Sample-size calculations use a variation of the type II error probability known as the power of the test. The **power** of a hypothesis test is the probability you'll correctly reject the null hypothesis, in other words, the probability you'll reject H_0 when it isn't true. It's one minus the type II error probability, or $1-\beta$. Sample size has a big impact on the power of a test, and if you gather enough samples, you can amass as much power as you like.

Delving into the world of statistical power is a little like delving into the world of political power: messy. For example, think of a simple test for the mean of a population,

$$H_0 : \mu = \mu_0 \text{ vs.}$$
$$H_A : \mu \neq \mu_0.$$

Notice the alternative hypothesis. If it were straightforward like $H_A : \mu = 5$, for example, I could, with enough statistical knowledge, assume my population had a mean of five, plug this mean value into the appropriate probability distribution, and calculate the probability my test statistic will fall below the critical level, leading me to incorrectly accept H_0. But I don't know how plug $\mu \neq 0$ into any probability distribution and get back a meaningful value. In other words, for one-sided and two-sided hypothesis tests, where the alternative hypothesis is $H_A : \mu < 0, \mu > 0$, or $\mu \neq 0$, the power isn't easy to find. So rather than trying to calculate such a convoluted probability, statisticians generally focus on a specific **effect size**, a practically meaningful difference between the H_0 mean and the actual mean. In other words, the effect size is the smallest, practically significant difference.

Take a t-test for example, where $H_0: \mu = \mu_0$ and $H_A: \mu > \mu_0$. The test statistic for this two-sided test is a t-statistic,

$$T = \sqrt{N} \frac{\bar{x} - \mu_0}{s}.$$

Here, $\bar{x}$ is the sample mean, s is the sample standard deviation, and N is the sample size. This test statistic depends on the sample size and the ratio $(\bar{x} - \mu_0) / s$. This ratio is a scaled difference, in other words, the difference between the sample mean and the H_0 mean relative to the standard deviation. The effect size is based on this type of scaled difference, where

$$\text{Effect size} = E = \frac{\mu - \mu_0}{s}.$$

For example, if the practically significant mean $\mu = 3$, the H_0 mean $\mu_0 = 1$, and $s = 1$, the effect size is $(3-1)/2 = 2$. This reflects a twofold difference in means, relative to the variation inherent in the population.

To calculate a sample size for a mean test, you specify a significance level α, a practically significant effect size E, and the power you'd like the test to have at that effect size. These three parameters give you everything you need to perform a sample-size calculation. Specifically, with a one-sided t-test having significance level α, you'd reject the null hypothesis if the test statistic T is larger than the corresponding critical value for the t-distribution with $N-1$ degrees of freedom. Mathematically, this can be written as follows:

$$T = \sqrt{N}\,\frac{\overline{x} - \mu_0}{s} > t_{1-\alpha, N-1}.$$

To achieve a specified power P ($P = 0.8$ is a common choice), you'd select N so that for effect size E, the probability of rejecting H_0 is P. Here's this probability in the form of an equation:

$$P\left\{\sqrt{N} \times E > t_{1-\alpha, N-1}\right\} = P$$

The idea is simple. Take this equation, plug in the desired significance level, effect size, and power, and find the sample size N that makes this equation hold true. Unfortunately, putting this idea into practice isn't as easy as performing a little algebra. Solving the above equation for N is complicated due to what's called the noncentral t-distribution, and so there are no simple formulas to calculate the sample size this way. There are, however, tables, functions in many data analysis packages, and free online sample-size calculators, and these tools can do the equation solving for you.

This same process can be used to calculate the sample size for virtually any hypothesis test. You specify the test statistic, plugging in a practically significant result and a statistical significance level, and state the criteria for rejecting the null hypothesis. Then you set the probability of that result to the desired power of the test and solve for the sample size. As with the t-test, however, the idea is much simpler than the implementation. For most common tests, solving these complicated probability equations can be difficult, and so finding a good data analysis software package that'll do these calculations for you is well worth the time and effort.

Steve's senatorial race is going beautifully. His new campaign platform—One People, One State—has almost become a state motto, and with his boyish good looks and easy smile he could sell matches to a woman trapped in a burning building. His poll numbers reflect this. The last poll of $N = 1064$ likely voters had him up 54% with an MOE of 2.9%. This could end up being one of the biggest victories in the state's history.

Just as the campaign enters its last crucial weeks before the November election, tragedy strikes. A compromising photo pops up on the Internet. Who posted the photo, we don't know, but the image is clear. It's a younger Steve, half-naked, driving a lawn tractor, waving an empty bottle of whiskey, and sporting a very questionable tattoo on his chest. Within minutes of being posted, this picture goes viral.

We begin damage control, searching for an explanation that can make this whole scandal go away. The public image consultants begin weaving together a story. A college mishap. No, a fraternity prank. Even better, a hazing ritual in which Steve was forced to participate if he wanted to belong to the cherished fraternity of his father and grandfather. It's perfect. It makes Steve more sympathetic and gives him a new cause: to stop bullying.

While the spin doctors work their magic, we decide to run a new poll, this one to determine just how much damage the photo caused. We plan to send our army of interns out to the voters to ask two questions: (i) if the person is aware of the tragic photo and (ii) if the person is considering voting for Steve in the upcoming election. This will give us two groups of voters to compare: those voters who haven't heard of the scandal and those voters who have. By using quota sampling, where we have the interns keep polling people until there's a specified number of likely voters in each of the two groups, we can make sure we get enough numbers in each sample to determine whether this unfortunate event impacts the peoples' opinion of our candidate.

If p_1 is the proportion of voters who favor Steve and do not know about the scandal, and if p_2 is the proportion of voters who favor Steve and do know about the scandal, then the following hypotheses can be tested:

$$H_0 : p_1 = p_2$$
$$H_A : p_1 > p_2$$

This is a one-sided hypothesis test for a proportion using two samples, and it can be run using a two sample z-test. If I use quota sampling and force the sample size of each group to be the same, or $N = N_1 = N_2$, then the test statistic is a z-statistic,

$$Z = \frac{\hat{p}_1 - \hat{p}_2}{\sqrt{\dfrac{2\hat{r}(1 - \hat{r})}{N}}}$$

where $\hat{r}$ is the average of $\hat{p}_1$ and $\hat{p}_1$, or $(\hat{p}_1 + \hat{p}_2)/2$. The null hypothesis is rejected if $Z > z_{1-a}$ where $z_{1-\alpha}$ is the $1-\alpha$ critical value for the standard normal distribution. This test statistic is the foundation of the sample size calculations for the new poll.

Many reputable statistical consulting firms and software companies have free online sample-size calculators. There's a documented, validated sample-size calculator at http://powerandsamplesize.com/ that can be used for a two sample proportion test. If I assume 5% points is a practically significant hit to Steve's image, this drops his 54% favorability rating down to 49%, or $\hat{p}_1 = 0.54$ and $\hat{p}_2 = 0.49$. Using these proportions, a significance level of $\alpha = 0.05$, a power of $P = 0.8$, and the *one-sided test* for H_0: $p_1 = p_2$ versus H_A: $p_1 > p_2$, the calculated sample size is $N = 1231$. In other words, I need to poll 1232 likely voters in each group if I want to be 80% sure I can detect a 5% point decrease in my candidate's favorability rating.

STEP 4. KEEP YOUR FRIENDS CLOSE AND YOUR ENEMIES CLOSER

After a little investigation, we determine the photo was posted by one of our interns, an attractive young woman who was once very loyal to Steve but, after an unfortunate misunderstanding in his office late one night, became angry and unhappy with the direction the campaign was going. She also didn't appreciate the amount of legwork involved in collecting so much poll data. Who knew a large sample size could become your own worst enemy?

You probably went away from your first statistics course thinking a large sample size is a good thing. After all, confidence intervals, hypothesis tests, regression, all these techniques are impacted by sample size, and as far as reducing uncertainty is concerned, a large N is your best friend. However, the 2012 presidential polls are a good example of how a bigger N doesn't necessarily translate into better predictions. In the days leading up to the election, the Gallup organization, one of the leading survey organizations in the country, polled 2700 voters and had Republican challenger Romney leading the race 50–49% with an MOE of 2%. The lesser known IBD/TTP organization, polled 712 voters and had the incumbent Obama leading the race 50–49%, with an MOE of 3.7%. Both polls had the candidates in a statistical dead heat. Gallup polled more people and had a smaller MOE, but in the end the IBD/TTP poll was closer to the final outcome of 51–47% in favor of Obama.

There are two reasons why a larger sample size doesn't always help you. First, a nice large N can lull you into a false sense of confidence in your results. But be warned. It's not enough to collect a lot of data, you also need to collect the right data. It's crucial to employ proper sampling strategies like those introduced in the Chapter 2, or you could find yourself with a very small MOE than means nothing because your sample doesn't reflect the population from which it was drawn. Second, large sample sizes produce very small uncertainties when it comes to confidence intervals and hypothesis tests. It's possible to make N so large, your statistical procedure rejects results as statistically significant when the actual difference is so small, it's practically insignificant.

A well-designed study applies proper sampling techniques in order to get a sample that adequately represents your population. It also incorporates just the right number of samples, determined from a combination of practical considerations and statistical sample-size calculations. In this way, you can amass statistical power, and for a modern campaign, where polls and image are everything, statistical power is political power.

BIBLIOGRAPHY

Barnett VIC. *Sample Survey Principles and Methods*. 3rd ed. London: Arnold; 2002.

Boerma L. Hurricane Sandy: Election 2012's October Surprise? Available at http://www.cbsnews.com/news/hurricane-sandy-election-2012s-october-surprise/. Accessed October 28, 2012.

CNN News Poll Methodology and Results. Available at http://political-ticker.blogs.cnn.com/2013/09/09/obama-hits-new-low-on-foreign-policy-in-cnn-polling/?hpt=po_c1. Accessed September 17, 2013.

FOX News. http://www.foxnews.com/politics/interactive/2013/09/09/fox-news-poll-voters-say-us-less-respected-since-obama-took-office/. Accessed September 17, 2013.

Frenz R. Types of Likert Scales. Available at http://www.ehow.com/list_6855863_types-likert-scales.html. Accessed May 14, 2012.

Good PI, Hardin JW. *Common Errors in Statistics (and How to Avoid Them)*. Hoboken: John Wiley & Sons, Inc; 2006.

Hall S. How to Determine the Sample Size of an Experiment. Available at http://www.ehow.com/how_6110532_determine-sample-size-experiment.html. Accessed May 30, 2014.

HyLown Consulting, LLC. Power and Sample Size Calculators. Available at http://powerandsamplesize.com/. Accessed August 20, 2014.

NIST/SEMATECH. Sample Sizes Required. e-Handbook of Statistical Methods. Available at http://www.itl.nist.gov/div898/handbook/prc/section2/prc242.htm. Accessed July 15, 2014.

Rasmussen S. Comparing approval ratings from different polling firms. Rasmussen Reports. March 17, 2009. Available at http://www.rasmussenreports.com/public_content/political_commentary/commentary_by_scott_rasmussen/comparing_approval_ratings_from_different_polling_firms. Accessed September 17, 2013.

RealClear Politics. General Election: Romney vs. Obama. Available at http://www.realclearpolitics.com/epolls/2012/president/us/general_election_romney_vs_obama-1171.html. Accessed May 30, 2014.

RealClear Politics. President Obama Job Approval, Real Clear Politics. Available at http://www.realclearpolitics.com/epolls/other/president_obama_job_approval-1044.html. Accessed September 17, 2013.

Schaeffer RL, Mendenhall III W, Lyman Ott R. *Elementary Survey Sampling*. 5th ed. Belmont: Duxbury Press; 1996.

Whitley E, Ball J. Statistics review 4: sample size calculations. *Crit Care* 2002;**6** (4):335–341.

4

BUNCO, BRICKS, AND MARKED CARDS: CHI-SQUARED TESTS AND HOW TO BEAT A CHEATER

Imagine you're in Las Vegas, in a dark room somewhere along the alleyways off the strip. A slick man in a leisure suit smiles at you from across a dimly lit table. You're playing a dice game called Sixes Bet. Ordinarily, you'd stay away from this game. After all, the odds of winning money at Sixes Bet are grim. But Mr. Slick has turned the tables, and based on your calculations, the probability you'll win each game is more than a half. The best odds you'll find anywhere in Sin City.

But if your odds are so good, why is your stack of betting chips getting so small?

Most basic statistical techniques work on continuous data, numeric observations that can take on any real value in some range. But not all observations are continuous, or even, for that matter, numeric. Sometimes you've got discrete data—whole numbers, integers, categories, or grouped observations. Analyzing such data takes something different than a sample mean, standard deviation, and t-test.

There are a relatively small number of basic techniques for analyzing discrete data, but these techniques are versatile and can answer many different types of questions. Like whether this greasy man sitting across from you, grinning with satisfaction, is really playing fair or not. This chapter presents common methods for discrete data analysis and shows how the chi-squared test can help you beat a cheater.

Beyond Basic Statistics: Tips, Tricks, and Techniques Every Data Analyst Should Know, First Edition. Kristin H. Jarman.
© 2015 John Wiley & Sons, Inc. Published 2015 by John Wiley & Sons, Inc.

WHAT HAPPENS IN VEGAS ... HOW STATISTICIANS REMAIN DISCRETE

The field of statistics is built on the concept of probability. A **probability** is a number between zero and one that expresses how likely some future event is to occur. A probability of zero means the event is simply not going to happen. A probability of one means it's a sure thing. When it's somewhere between zero and one, you have an event whose outcome is uncertain. For example, the probability there's gambling happening in Las Vegas at this very moment is one. The probability every casino has suddenly and unexpectedly run out of booze is zero. The probability you'll get food poisoning next time you visit one of the all-you-can-eat buffets is somewhere between zero and one.

Most people are comfortable with probabilities, as long as they stay outside the realm of mathematics. For example, if you've ever answered a question with the words, "probably," "not likely," or "never going to happen" you've used the concept of probability to measure the likelihood of some event. However, to the data analyst, probability is a much more precise, mathematical term that's used everywhere statistics is used. If you're reading this book, you're probably familiar with the concept of probability. Maybe you've had a basic statistics course. Maybe you've tossed coins in class to understand how mathematics can be used to describe uncertain outcomes. Maybe you've spent time in Vegas, gambling, watching the laws of probability boost, or more likely, drain your bank account. If any of these are you (and I promise, I'm not telling), feel free to skip to the next section. On the other hand, if you're in need of a brief refresher, this section will provide you with the basics you need to get through the rest of the chapter.

To construct a formal, mathematical probability, you need three things: (i) a **random experiment**, a future trial, game, situation, or circumstance whose outcome is not yet known, (ii) an **event**, or outcome you'd like to see, and (iii) some way to assign a value to the likelihood of that event. For example, the roll of a six-sided fair die is a random experiment. The number "5" might be an event, or outcome you'd like to see. There are six sides to the die, all equally likely to appear, and so the probability of rolling a five is one out of six possible outcomes. More formally, if R represents the roll of your die, then $P\{R=5\} = 1/6$.

Any random experiment produces outcomes, and these outcomes, which are also called observations or measurements, can be either continuous or discrete. Continuous outcomes are the focus of the remaining chapters in this book. Discrete observations are the focus of this one.

Discrete observations are whole numbers, counts, or categories, in other words, anything that can be listed. For example, if you roll a six-sided die, you'll get a 1, 2, 3, 4, 5, or 6. If you roll two dice and add the faces together, you'll get a 2, 3, 4, 5, 6, 7, 8, 9, 10, 11, or 12. If you roll 400 dice and add the faces together, you'll get a whole number between 400 and 2400. It may not be fun, but it's possible to list each and every outcome in all of these experiments.

Simple discrete probabilities, like the probability a coin toss will land on *heads*, can be calculated without any sophisticated math. There are two sides to a coin, one of which is *heads*. The probability of a coin landing on *heads* is one out of two possible outcomes, or 1/2. Unfortunately, life is rarely this simple. Data analysts are usually interested in more complicated outcomes, for example, the probability of a coin landing on heads five times in a row. Or the probability eight out of the next ten coin tosses will land on heads. Trying to reason your way through probabilities such as these quickly becomes unmanageable. That's why we have random variables and probability distributions.

A **random variable** represents the outcome of a random experiment. Typically denoted by a capital letter such as X or Y, a random variable takes on the value some as-yet-undetermined outcome of a random experiment. For example, on a coin toss, with possible outcomes *heads* and *tails*, you could define a random variable $X=0$ for *tails* and $X=1$ for *heads*. This value of X is undetermined until the experiment is complete.

A **probability distribution** is a mathematical formula for assigning probabilities to the values of a random variable. Many different probability distributions have been developed over the years, and these can be used to assign probabilities in virtually any random experiment. **Discrete probability distributions** describe random experiments whose possible outcomes are discrete. To qualify as a discrete probability distribution, a function must have a probability associated with every possible outcome. Each probability must be a value between zero and one. And if you add the probabilities of all possible outcomes together, the answer should be one. There are many common discrete probability distributions. The simplest is known as the Bernoulli distribution. If you have a random experiment with two possible outcomes, the Bernoulli distribution assigns probabilities to both. Here's the formula:

$$P\{X = x\} = \begin{cases} p & \text{when} \quad x = 0 \\ 1 - p & \text{when} \quad x = 1 \end{cases}$$

Think of a coin toss. The outcome of that coin toss could be represented by a random variable X with $X=0$ for *tails* and $X=1$ for *heads*. If the coin is fair, the probability of tails is $P\{X=0\} = 1/2$ and the probability of heads is $P\{X=1\} = 1/2$. If the coin is unevenly weighted so that it tends to land on heads, then you might have $P\{X=0\} = 1/3$ and $P\{X=1\} = 2/3$. Both of these

Distribution name	Parameters	What it measures	The mean and variance
Bernoulli	Success probability p	Any random experiment with two outcomes	$\mu=p$ $\sigma^2=p(1-p)$
Binomial	Number of trials N Success probability p	The number of successes in N independent trials, or the sum of N independent Bernoulli random variables	$\mu=Np$ $\sigma^2=Np(1-p)$
Geometric	Success probability p	The number of independent trials needed until the first success	$\mu=1/p$ $\sigma^2=(1-p)/p^2$
Negative binomial	Success probability p Number of failures r	In independent trials, the number of successes that occur before the rth failure occurs	$\mu=pr/(1-p)$ $\sigma^2=pr/(1-p)^2$
Poisson	Success rate λ per unit time Success rate λ per unit distance or area	The number of successes per unit time (or distance/area)	$\mu=\lambda$ $\sigma^2=\lambda$

FIGURE 4.1　Common discrete probability distributions.

are instances of the Bernoulli distribution—one with success probability $p=1/2$ and the other with success probability $p=2/3$.

The Bernoulli distribution is the simplest example of a discrete probability distribution. There are many more that can be used to describe more complicated random experiments. Figure 4.1 lists some of the most common ones.

CONTINGENCY TABLES, CHI-SQUARED TESTS, AND OTHER WINNING STRATEGIES FOR DISCRETE DATA ANALYSIS

Many types of studies produce discrete data. For example, surveys commonly ask for personal information such as gender, race, and education level. Internet search engine companies and online retailers use past search history – strings of words – to tailor searches, recommend products, and target popup ads to an individual. Market research companies track sales trends by looking at who bought what product and why. Demographic data, text, and personal preference, all of these fall into the category of discrete data.

There are two types of discrete data: qualitative and quantitative. **Discrete qualitative data** consists of observations that cannot be ordered in any mathematically meaningful way. A typical Las Vegas slot machine produces discrete qualitative data. There are three wheels on the machine, and each wheel has pictures of different items: cherries, lemons, gold bars, and so on. When you place your bet and pull the lever (or push the button), the three wheels spin for a while, and each eventually stops on one of these items. Every item is distinct, but there's no way to compare them mathematically. A cherry isn't more or less than a lemon. It's just different.

Discrete quantitative data is discrete data that has a meaningful mathematical order to it. For example, the game of craps involves betting on the sum of a roll of two dice. The outcome is a number between two and twelve. A roll of twelve is more than a roll of eleven, which is more than a roll of ten, and so on. In other words, these outcomes can be mathematically, or quantitatively, compared.

In many ways, discrete quantitative data is a lot like continuous data. Because it's numeric, you can calculate a meaningful sample mean and standard deviation. And thanks to important results like the central limit theorem and the normal approximation to the binomial distribution, you can apply popular techniques such as *t*-tests and analysis of variance. Discrete qualitative data can be a bit trickier to analyze – it's not numeric and so there aren't as many statistical techniques available. This section describes some methods for analyzing all discrete data, but you'll find them particularly useful for qualitative data.

Turning Lemons into Gold Bars: How to Convert Qualitative Data into Quantitative Random Variables

Suppose you want to calculate the probability your next play of a $1 slot machine will hit the jackpot. You need to land on three gold bars to hit this jackpot. A typical modern Vegas slot machine might have 256 items on each wheel. The probability that any given wheel will land on a gold bar is one chance out of 256, or 1/256. Using a probability rule called the multiplication rule, the probability all three wheels will land on a gold bar is $1/256 \times 1/256 \times 1/256 = 0.00000006$.

You don't need to be an expert in probability theory to realize this outcome is extremely unlikely. If you're counting on these winnings to buy a plane ticket home, you might as well just keep your dollar and apply for a job washing dishes instead. On the other hand, you don't need three gold bars to win money. There are other outcomes that pay out as well. For example, if you get at least two of a higher ranking item, the cherries maybe, you don't win the jackpot, but you do win some money. And if you allow yourself to play a number of games, you increase your chances of winning at least once over the long run.

You can turn lemons into gold bars and increase your chance of success by focusing on a more realistic slot machine strategy. There are many ways to do this. For example, if you restrict yourself to thirty games, what's the probability you'll win more games than you'll lose? What's the probability you'll walk away with a little extra cash? How many free drinks will the average cocktail waitress bring you before all your money is gone? And does the savings on all those free drinks make up for any losses at the slots?

With questions like these, it's often possible to construct a random variable that transforms qualitative data into a quantitative random variable. This technique, based on the Bernoulli distribution, is usually taught in basic probability courses, but its usefulness can be easy to overlook. Recall, a Bernoulli random variable is one that takes one of two possible values, zero for failure and one for success. By cleverly defining what it means for an outcome to be a success, you can define a Bernoulli random variable that can be used to calculate any number of more complicated probabilities. For example, suppose you've gone through all the possible wheel combinations and scoured the Internet, and you've calculated the probability of winning at least \$1 (your bet plus some money) on the slot machines to be 0.3. If you define a random variable X to be zero if you lose your dollar a single game and one if you break even or win, then the probability distribution becomes

$$P\{X = x\} = \begin{cases} 0.7 & \text{when } x = 0, \quad \text{or you lose} \\ 0.3 & \text{when } x = 1, \quad \text{or you win.} \end{cases}$$

The outcome of successive games can be described by probability distributions for independent trials. **Independent trials** are repeated random experiments where the outcome of the any one trial is not impacted by the outcome of any previous trials. Playing thirty games of slots is an example of thirty independent trials. The probability you win on the ninth roll is 0.3, completely unaffected by how many games you played or won before it.

There are many basic probability distributions that apply to independent trials, and the most common of these are listed in Figure 4.1. For example, the **binomial probability distribution** describes the number of successes in a given number of independent trials. Mathematically, if you have N independent trials, each with probability of success $P\{X=1\}=p$, and if Y represents the number of successes, in other words, the sum of N Bernoulli random variables, then

$$P\{Y = i\} = \binom{N}{i} p^i (1 - p)^{N-i}$$

for any $i=0, 1, 2, \ldots, N$. This is the binomial distribution. For example, if Y is the number of games you've won out of thirty total, then Y follows the binomial probability distribution with $N=30$ and $p=0.3$.

You can use the binomial distribution to calculate the average number of games you expect to win out of thirty total. You can calculate the probability you'll win at least half of the games. You can even use an approximation called the normal approximation to the binomial distribution and apply common techniques for continuous data. For further details on how to work with the binomial distribution, I refer you to a basic statistics text such as *The Art of Data Analysis: How to Answer Almost Any Question Using Basic Statistics.*

Plots and Tables: The Poor Man's Statistical Analysis

Like the tourist who went to Vegas with his pockets full of cash and his head full of ideas, a data analyst can't always make his quantitative data dreams come true. Sometimes, you're stuck with qualitative data. One of the best ways to analyze this type of data is through plots and contingency tables.

It's always important to get a good look at your data. The right graph can help you spot trends and outliers. It can highlight problems such as data entry errors. In some cases, the right graph can even eliminate the need for further statistical analysis. The most popular type of graph for discrete qualitative data is the frequency distribution, plotted as a simple bar chart. A **frequency distribution** is a tally of the number of times each different category or value appears in a discrete dataset. The frequency values can be expressed as whole numbers, relative frequencies (fractions of the total number), or percentages (the relative frequencies multiplied by 100). For example, suppose I work for the Nevada Bureau of Tourism, and we've just come up with a new ad campaign to entice more visitors to come to Vegas. Before we launch this campaign, I'd like to test what its impact might be. Using a five-point Likert scale (see Chapter 2), I put together a survey. One of the questions on the survey is as follows:

How likely are you to visit Las Vegas in the next twelve months?

1. Highly unlikely
2. Unlikely
3. Neutral
4. Likely
5. Highly likely

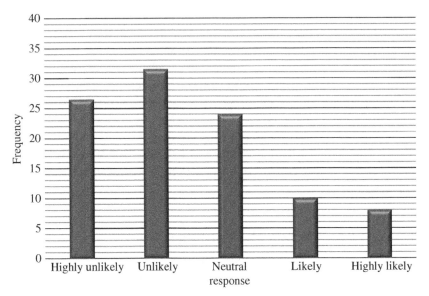

FIGURE 4.2   How likely are you to visit Vegas in the next twelve months?

I gather a group of 200 frequent travelers and ask them this question. Here are the results (Figure 4.2).

With one glance, it's easy to see that the bars are skewed in the negative direction. In fact, the most popular response, the mode, is "unlikely." Over 30% of respondents said they'd be unlikely to visit Vegas in the next year. Added together, the "unlikely" and "highly unlikely" categories make up more than 50% of all the responses.

In order to measure the impact of our new ad campaign, I then take these same volunteers and expose them to all our materials—commercials, brochures, Internet pop-ups, the whole bit. Afterward, I once again ask them how likely they'd be to visit Vegas in the next year. This type of before-and-after approach, called a repeated measures study, gives me a direct way to observe the effect these ads have on the traveling public.

A bar chart can be used compare the before- and after-data. Figure 4.3 plots such a bar chart, where the "before" frequencies for each response are plotted as darker bars on the left, while "after" frequencies are plotted in a lighter shade on the right. The results show a definite shift in the positive direction. The mode, or most common response, is now "Neutral." Where fewer than 20% of respondents reported being likely or highly likely to visit Vegas before seeing the ad campaign, now 30% reported the same. This 10% shift in the positive direction could mean millions in tourist dollars flowing in every year.

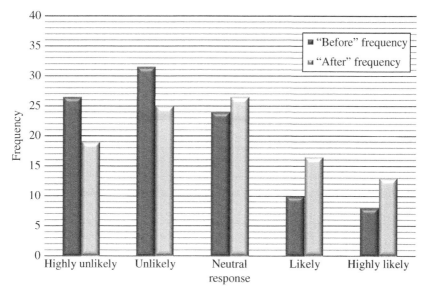

FIGURE 4.3 Before and after frequencies for Las Vegas survey.

Contingency Tables: How to Break Down a Frequency Distribution and Expose Your Variables

Suppose the Vegas ad campaign features show girls, scantily-dressed women dancing on stage with props that I'll leave to your imagination. With visuals like these, I'd expect the average man to respond positively to the ad campaign. But this type of blatant marketing might backfire if women don't like what they see. In this case, it's useful to look at my data by gender and ask the question, "How are men and women influenced by the ad campaign?"

A **contingency table** is a tool for breaking down a frequency distribution by variables. In a contingency table, the frequencies are listed by variable *and* by outcome: variable by rows and outcome by columns. For example, a contingency table for the Vegas ad campaign data would have gender listed by row, and likelihood of visiting the city by column. The number of survey respondents fitting each variable/outcome combination would then be listed in the appropriate cell of the table. Figure 4.4 shows a contingency table containing the test group's responses before watching the ads.

This table sums up the results nicely. It shows the number of men and women who responded to the survey question in every possible way. The far right column adds up the number of responses in each category for men and women, giving a total number of men (100) and women (100) who participated in the study. The bottom row sums up the total number of responses in each category, giving the "before" frequency distribution plotted in Figure 4.3.

Gender	How likely are you to visit in the next twelve months?					
	Highly nlikely	Unlikely	Neutral	Likely	Highly likely	Total
Men	26	31	26	10	7	100
Women	27	32	22	10	9	100
Total	53	63	48	20	16	200

FIGURE 4.4   Two-way contingency table for Vegas ad campaign *before* watching ads.

Gender	How likely are you to visit in the next twelve months?					
	Highly nlikely	Unlikely	Neutral	Likely	Highly likely	Total
Men	15	20	27	21	17	100
Women	23	30	26	12	9	100
Total	38	50	53	33	26	200

FIGURE 4.5   Two-way contingency table for Vegas ad campaign *after* watching ads.

Does my racy ad campaign affect men differently from women when it comes to their view of Las Vegas? Figure 4.5 helps answer this question. This figure displays a contingency table of men's and women's responses after being exposed. To the ads, that is.

According to the tables in Figures 4.4 and 4.5, women don't appear to be negatively impacted by the ad campaign. The number of women in the *highly unlikely* and *unlikely* categories drops a little after watching the ads, while the number in the *neutral* category increases. The *likely* and *highly likely* categories remain essentially unchanged. Men, on the other hand, appear to be more strongly impacted. In their case, the *likely* and *highly likely* categories more than double, while the negative responses drop noticeably.

Figures 4.4 and 4.5 are what we call **2×5 contingency tables**. They each have one variable with two possible values and one outcome with five possible values. The simplest and most common type of contingency table is a **2×2 contingency table**, having one variable with two possible values and one outcome with two possible values. Both types of contingency tables are

two-way contingency tables, meaning they look at one variable and one outcome. Both the variable and the outcome can have as many possible values as you like, but as long as you're looking at only a single variable/outcome combination, it's still a two-way contingency table.

A three-way contingency table looks a lot like a two-way contingency table, except the frequencies are broken down by two variables instead of just one. For example, suppose I'd like to break down my survey responses not only by gender but also by age. In this case, I'd add one layer to the table representing age ranges, for example, 20–29, 30–39, 40–49, 50–59, and 60+. This additional layer generates a contingency table like the one shown in Figure 4.6.

In this table, the number of responses for each gender/age combination are given. The column on the right adds up the total number of men and women in each age range, giving a frequency distribution of the respondents by gender and age. The row across the bottom adds up the total number of responses in each category, once again giving the frequency distribution shown in Figure 4.3.

Because a three-way contingency table breaks down the frequencies by two variables instead of just one, the counts in each cell are smaller than they were in the two-way contingency table. This will always happen when you increase the number of layers, or number of variable/outcome combinations,

Gender	Age	How likely are you to visit in the next twelve months?					
		Highly nlikely	Unlikely	Neutral	Likely	Highly likely	Total
Men	20–29	3	4	6	4	3	20
	30–39	5	7	8	5	5	30
	40–49	4	5	9	7	6	31
	50+	3	4	4	5	3	19
Women	20–29	4	4	3	4	4	19
	30–39	9	10	11	3	2	35
	40–49	8	10	9	2	1	30
	50+	2	6	3	3	2	16
Total		38	50	53	33	26	200

FIGURE 4.6 Three-way contingency table for Vegas ad campaign *after* watching ads.

represented in a contingency table. It isn't a problem when you have a large dataset with high frequencies in each cell, but when the counts start to fall below ten, certain statistical techniques for discrete data start to break down. This topic will be discussed more in the following section.

Higher order contingency tables with many layers do exist, but these can be difficult to tabulate and impossible to interpret, so when higher order contingency tables are needed, I usually revert to a graph such as the side-by-side frequency distributions plotted in Figure 4.3.

The Chi-Squared Test: An All-You-Can-Eat Buffet for Discrete Data Analysis

Bar charts and contingency tables are a great way to get a good look at your data, and when the outcome of a study is simple and clear cut, it may be all you need. But a more sophisticated statistical analysis can help you make conclusions when the data aren't so obvious. For example, Figures 4.4 and 4.5 show the Las Vegas responses before and after participants were exposed to my new marketing materials. There appears to be a difference between how men and women responded to the ads, but is it a significant difference? In other words, are the differences more than you'd expect purely by chance?

In my experience, there are three questions people commonly ask when analyzing studies involving discrete data, and this is one of them. Here's a list of all the three:

1. Is there any difference between categories, or is the frequency distribution evenly distributed across outcomes?
2. For before- and after-studies, did the treatment change the frequencies?
3. Does the value of one variable impact the frequencies for another? In other words, are any two variables correlated with one another?

All of these three questions can be answered using a hypothesis test. For example, suppose I want to know if frequent travelers have strong preconceived ideas about visiting Las Vegas. Without knowing any different, I might expect the responses to the survey question, "How likely are you to visit Las Vegas in the next twelve months?" to be pretty much evenly distributed between the five possible answers on the survey. In other words, the frequencies across all possible responses for this question would be about the same. Mathematically, suppose O_i represents the number of responses falling into category i, where the categories are *highly unlikely*, *unlikely*, *neutral*, *likely*, and *highly likely*. If there's no preference for one category over another, this number O_i would be approximately the same for all i. And if $N=200$ is the total number of respondents in the survey and $M=5$ is the number of categories, I'd expect it to be $N/M=200/5=40$ responses per category. Of course, real-world data are never perfect, so there will be differences between the O_i. The question is, are they statistically significant?

To answer this question, I'll set up the following two hypotheses:

$$H_0 : O_i = N / M \text{ for every response category } i \quad \text{vs.}$$
$$H_A : O_i \neq N / M \text{ for at least one response category } i.$$

Recall H_0 is the null hypothesis. It's the fall-back position, what you're automatically assuming to be true. H_A is the alternative hypothesis. This is the claim you accept as true only if you have enough evidence in the data to reject H_0. The test for these two hypotheses can be performed using a **chi-squared** test. When it comes to hypothesis testing, the chi-squared test is an all-you-can-eat buffet for data analysts. There's something to appeal to everyone, whether you've got discrete or continuous data. In fact, it's the only method I know that, with little to no modification, can test for several completely different hypotheses. The rest of this section illustrates just how versatile this one little test is.

Goodness-of-Fit Tests   The test for hypotheses

$$H_0 : O_i = N / M \text{ for every response category } i \text{ vs.}$$
$$H_A : O_i \neq N / M \text{ for at least one response category } i$$

is a test for the underlying probability distribution of a population. The observed frequency in each category is compared to the frequency you'd expect if the underlying probability distribution were a **discrete uniform distribution**, where the probability of an observation falling into each category is the same across categories. Any hypothesis test for the underlying probability distribution of a population is called a **goodness of fit**.

The most common goodness-of-fit test compares a sample of discrete observations to the discrete uniform distribution, but it doesn't need to be this way. You can construct a goodness-of-fit test for any probability distribution you can imagine. For example, even if people have no preconceived notions about visiting Las Vegas, I wouldn't necessarily expect the survey responses to be uniformly spread out between all of the categories. I might expect most of them to fall into the *neutral* category, with fewer responses on the positive and negative side. In this case, I could construct a different underlying probability distribution, one whose highest probability is the *neutral* category, and run my goodness-of-fit test against this distribution.

The standard goodness-of-fit test works on counts, or frequencies, not probabilities. Here are the hypotheses:

$$H_0 : O_i = E_i \text{ for every category } i \quad \text{vs.}$$
$$H_A : O_i \neq E_i \text{ for at least one category } i.$$

The value O_i is the observed number of counts in category i. The value E_i is the number you'd expect *if your population followed the probability distribution you're testing against*. This **expected frequency** is calculated by multiplying the probability of each category by the total number of samples. For example, in testing the Vegas survey against a discrete uniform distribution, there are five possible responses. Under the null hypothesis, the probability of a response falling into any category is 1/5. The total number in my sample is 200. So, the expected frequency is $200 \times 1/5 = 40$ responses per category.

In general, for a population with M outcome categories, the test statistic for the goodness-of-fit test is

$$X^2 = \frac{(O_1 - E_1)^2}{E_1} + \frac{(O_2 - E_2)^2}{E_2} + \cdots + \frac{(O_M - E_M)^2}{E_M}$$

There are two important requirements for the expected frequencies. First, these probabilities can come from a known probability distribution or they can be values you specify—anything that makes sense—but they must form a proper probability distribution. In other words, there must be a nonzero probability for every category, and they must add to one. Second, the expected frequencies in each cell should be at least five. In other words, you should have a large enough sample so that $E_i \geq 5$ for every i.

If, by some odd circumstance, the observed frequencies in each category exactly match the expected frequencies, the test statistic X^2 will be zero. The more the observed frequencies deviate from the expected frequencies, the larger X^2 will be. If this test statistic is large enough, larger than you'd expect by random chance alone, then you can declare statistical significance and conclude your population does not conform to the specified probability distribution. In other words, you compare the test statistic X^2 to a decision threshold. This threshold, $\chi^2_{1-\alpha, M-1}$, is the $1-\alpha$ critical value for the chi-squared distribution with $M-1$ degrees of freedom, where M is the number of categories, not the number of samples. If the test statistic is greater than this threshold, you reject H_0.

Most data analysis packages have a chi-squared test built into them, and these procedures can perform all the calculations for you. However, it's useful to work through an example at least once. To that end, the observed

How likely are you to visit Vegas within twelve months?	"Before" frequency	"After" frequency
Highly unlikely	43	38
Unlikely	63	50
Neutral	52	53
Likely	25	33
Highly likely	17	26

FIGURE 4.7   Frequency distribution for Vegas tourism survey.

frequencies for the Las Vegas survey responses are shown in Figure 4.7. Under a discrete uniform distribution, the expected frequencies are $E_i = 40$ per category. To determine if the "Before" frequencies are consistent with a discrete uniform distribution, the test statistic is

$$X^2 = \frac{(43-40)^2}{40} + \frac{(63-40)^2}{40} + \frac{(52-40)^2}{40} + \frac{(25-40)^2}{40} + \frac{(17-40)^2}{40} = 43.5$$

Critical values for the chi-squared distribution are provided in Appendix C. For the Las Vegas survey question, $M = 5$. For $\alpha = 0.05$, the critical value, or decision threshold, is $\chi^2_{0.95,4} = 9.5$. Since the test statistic, $X^2 = 43.5$, is larger than this decision threshold, I reject the null hypothesis that the survey responses are uniformly distributed across the different categories. In other words, I conclude at least one of the categories has significantly more or fewer responses than I'd expect if they were all equally likely.

Recall, with a "≠" in the alternative hypothesis, this test is a two-sided hypothesis test. Deviations from the null hypothesis in one direction or another are equally significant. There is a one-sided version of the goodness-of-fit test, where the alternative hypothesis specifies either greater than or less than. However, it's messy, and I've found I can usually do without it. If you need a one-sided goodness-of-fit test, I refer you to *Nonparametric Statistical Methods* by Hollander and Wolfe for all the details.

The Chi-Squared Test for Contingency Tables   The idea behind a goodness-of-fit test is simple. Compare observed frequencies to expected frequencies and determine if the difference between them is statistically significant. This idea can be applied to various situations. Contingency tables, for example, can be analyzed using a chi-squared test. The procedure is straightforward, as long as you understand how to calculate the expected frequencies you need for the test.

Think of a two-way contingency table, with one variable and one outcome. The chi-squared test for this type of table is

$$H_0: O_{ij} = E_{ij} \text{ for variable } i \text{ and category } j \quad \text{vs.}$$

$$H_A: O_{ij} \neq E_{ij} \text{ for at least one variable } i \text{ / category } j \text{ combination.}$$

The chi-squared test statistic is the same as before, namely a sum of squared differences between the observed and expected frequencies. But now you have two indices, i and j, instead of one. This makes the test statistic slightly more complicated. If your variable has L possible values and your categories have M possible values, the test statistic for this test is

$$X^2 = \frac{(O_{11} - E_{11})^2}{E_{11}} + \frac{(O_{12} - E_{12})^2}{E_{12}} + \cdots + \frac{(O_{1M} - E_{1M})^2}{E_{1M}}$$

$$+ \frac{(O_{21} - E_{21})^2}{E_{21}} + \frac{(O_{22} - E_{22})^2}{E_{22}} + \cdots + \frac{(O_{2M} - E_{2M})^2}{E_{2M}}$$

$$\cdots$$

$$+ \frac{(O_{L1} - E_{L1})^2}{E_{L1}} + \frac{(O_{L2} - E_{L2})^2}{E_{L2}} + \cdots + \frac{(O_{LM} - E_{LM})^2}{E_{LM}}$$

The observed frequencies O_{ij} are just the counts in each cell of the contingency table: simple. The expected frequencies E_{ij} are the values you'd expect if your data conform to a distribution determined by probabilities p_{ij}: not quite so simple. Calculating this requires either (i) a formula or (ii) careful thought. I'll give you the formula below.

A two-way contingency table includes one variable and one type of outcome. Both the variable and the outcome can have as many possible values as you like. The expected frequency for variable i and outcome j, call it E_{ij}, is the number of observations having variable value i, N_i, multiplied by the probability of outcome j *under the null hypothesis*, p_j, or

$$E_{ij} = N_i \times p_j.$$

For example, the expected number of men who respond *highly unlikely* in the survey is the total number of men, N_1, times the probability of highly unlikely, p_1. Under the null hypothesis, this probability follows a uniform distribution where $p\_i = 1/5$ for all categories, and so this value is $E_{11} = 100 \times 1/5 = 20$. Similarly, the expected frequency of women who respond *neutral* is the number of women, N_2 times the probability of neutral, or $E_{23} = 100 \times 1/5 = 20$.

For the Vegas survey example, the number of men and women is exactly the same. And if I use a discrete uniform distribution for the null hypothesis, the probability of a response in each category is exactly the same. So, the expected frequencies in this example are all the same, twenty responses per gender and response category.

The decision threshold for the goodness-of-fit test in a two-way contingency table is the critical value for the chi-squared distribution with $(L-1) \times (M-1)$ degrees of freedom. Referring to Appendix C, with $L=2$ and $M=5$, the 0.05 chi-squared critical value with 4 degrees of freedom is $\chi^2_{0.95,4} = 9.5$.

With these expected frequencies and a decision threshold, testing the contingency tables in Figures 4.4 and 4.5 for goodness of fit against a discrete uniform distribution is straightforward. For example, the test for the before-frequencies gives a chi-square value of

$$X^2 = \frac{(26-20)^2}{20} + \frac{(31-20)^2}{20} + \cdots + \frac{(9-20)^2}{20} = 44.$$

Since this test statistic exceeds the decision threshold value 9.5, I'd reject the null hypothesis that the responses are uniformly distributed across all categories.

Testing for Independence   The frequencies in Figure 4.4 are about the same for men and women, suggesting that both sexes are similarly predisposed to plan a trip to Las Vegas. But there are differences. Are these differences statistically significant? In other words, are the outcome frequencies somehow related to the gender of the respondent? The chi-squared test, powerful little technique that it is, can be used to answer this question, too.

Statistical dependence refers to a relationship between two variables—whether categorical, discrete, or continuous. Technically, two random variables are **dependent** if the value of one impacts the probability of the outcome of the other. For example, suppose men are more inclined to want to visit Las Vegas than women. If you chose a man and a woman at random and asked both of them if they wanted to visit Vegas, the man would be more likely to say "yes" than the woman. This wouldn't be true for all men and women. Rather, the probability the man would say "yes" would be higher than probability for the woman. In this case, gender and the desire to visit Vegas are dependent variables.

When two variables a not dependent on one another, they are independent. Specifically, A and B are **independent** if

$$P\{A = a \text{ and } B = b\} = P\{A = a\} P\{B = b\}.$$

This means the value of one does not impact the probability of the value of the other, no matter what those values are. For example, suppose A represents gender and B represents a desire to visit Las Vegas. If gender and preference for visiting Vegas are statistically independent, then

$$P\{A = man \text{ and } B = highly \text{ likely}\} = P\{A = man\} \times P\{B = highly \text{ likely}\}$$

The same relationship is true for A=woman and B=highly likely, A=man and B=neutral, A=women and B=unlikely, and any other gender/outcome combination in the study.

The **chi-squared test for independence** can be used to determine if a variable and an outcome are independent, and it relies on this mathematical definition of independence. Specifically, if r_i is the probability of observing variable value i and p_j is the probability of observing outcome category j, then the probability of observing both variable value i and outcome category j is $p_{ij} = r_i \times p_j$. Under this scenario, the expected frequency in cell i,j of the corresponding contingency table is given by the relationship:

$$E_{ij} = N \times r_i \times p_j$$

Figure 4.8 shows how the contingency table can be used to calculate the expected frequencies for the Las Vegas survey data originally presented in Figure 4.4.

| Gender | How likely are you to visit in the next twelve months? | | | | | |
	Highly nlikely	Unlikely	Neutral	Likely	Highly likely	Total
Men	26	31	26	10	7	100 $\rightarrow r_1 = \# Men/N = 0.5$
Women	27	32	22	10	9	100
Total	53	63	48	20	16	200 (N)

$p_3 = \# \text{Neutral}/N = 0.54$

Expected frequency: $E_{13} = N \times r_1 \times p_3 = 100 \times 0.5 \times 0.54 = 27$
Observed frequency: $O_{13} = 26$

FIGURE 4.8 Calculating chi-squared frequencies for the Las Vegas ad campaign data.

The rest of the process is exactly the same as the original test for goodness of fit. Set up the hypotheses:

$$H_0 : O_{ij} = E_{ij} \text{ for every variable value } i \text{ and outcome } j \text{ vs.}$$

$$H_A : O_{ij} \neq E_{ij} \text{ for at least one } i \,/\, j \text{ combination}$$

Calculate the test statistic

$$X^2 = \frac{\left(O_{11} - E_{11}\right)^2}{E_{11}} + \frac{\left(O_{12} - E_{12}\right)^2}{E_{12}} + \cdots + \frac{\left(O_{LM} - E_{LM}\right)^2}{E_{LM}}$$

Under the null hypothesis, X^2 follows the chi-squared distribution with $(L-1) \times (M-1)$ degrees of freedom. The $1 - \alpha = 0.95$ critical value for this distribution is given by 9.5, so if X^2 exceeds this value, then the null hypothesis is rejected and I conclude the variable and outcomes are dependent.

For the data in Figure 4.8, $X^2 = 8.61$. This is less than the critical value of 9.48, so I'd accept the null hypothesis that gender and response are independent. In other words, I'd conclude that men and women have the same preconceived notions about visiting Vegas.

HOW TO BEAT A CHEATER

Sixes Bet is a simple dice game. You place your bet. The dealer rolls a single die four times. If he rolls no sixes, you win. Otherwise, you lose.

The probability you win a game of Sixes Bet is straightforward to calculate if you remember how to use the binomial distribution. Four rolls of the die, meaning four trials. The probability of a success—that's a failure to you—on each roll is the probability the dealer rolls a six, or 1/6. The probability you'll win a single game of Sixes Bet is the probability of zero sixes out of four rolls. This comes to

$$P\{\text{no sixes out of four rolls}\} = 0.48.$$

Sixes Bet is well known in Las Vegas. The probability of winning isn't bad, but when you do win, the payout is poor, and many unsuspecting gamblers have found themselves out of money quickly. You know this. Mr. Slick knows this. Mr. Slick knows you know this. So, like any good Vegas-style businessman, Mr. Slick has a hook. He's turned the tables on this game. He lets you play the dealer.

Since you're the one trying to roll at least one six, the probability you'll win each game is now

$$P\{\text{at least one six out of four rolls}\} = 0.52.$$

That's better odds than you'll find anywhere in Las Vegas.

As long as you're playing with a fair die, that is.

A fair die is a perfectly symmetrical and balanced cube, meaning the edge angles and weight distribution along every side are exactly the same. It's this perfect balance that creates equal probabilities for each of the six faces. Loaded dice are made by upsetting this balance. A quick Internet search reveals a number of ways to make a loaded die, for example, by shaving one face of an ordinary die ever so slightly, or by inserting a tiny nail in one face of the die to tip the odds in favor of a certain roll. The odds of you winning this version of Sixes Bet are good, and yet, you're losing money hand over fist. Could Mr. Slick be using a loaded die?

To answer this question, I recreated the scenario in the form of a **double-blind study**. I bought a set of four dice off the Internet. Two of the dice are fair. Two of them are loaded for the game of craps, meaning they have a slightly higher than normal probability of rolling a two or five. All four dice look exactly the same. I found a test subject and had her pick one of the dice at random. Then I asked her to play forty games of Sixes Bet according to Mr. Slick's rules. Neither my test subject nor I knew if the die she chose was loaded or not. In other words, both of us were blind to the true nature of die being tested. Here are the results of the experiment.

In all, my test subject won ten of the forty games. This is suspicious. With a fair die, the probability of winning each game is just over a half, so I'd expect her to have won about half of the games. Suspicious as this result may be, however, it isn't conclusive. Losing streaks do happen, after all, and I need a stronger argument if I'm going to beat this cheater.

To better understand what was happening with this die, I took these forty games and broke them out into the individual rolls, all 160 of them. Figure 4.9 plots the frequency distribution for these rolls. The frequencies are given above the bars so that a numerical comparison can be made.

This figure only adds to my suspicion. With a fair die, all of the numbers should have about the same frequency. However, there were only eleven sixes out of the 160 rolls. That's an empirical probability of 0.07, less than half what it should be for a fair die. On the other hand, the number two was rolled thirty-nine times. That's an empirical probability of 0.25, quite a bit

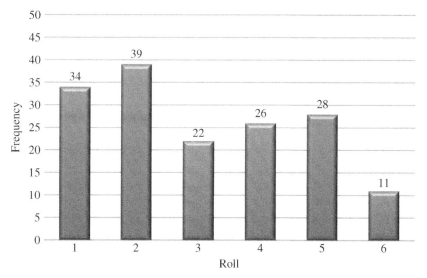

FIGURE 4.9 Frequency distribution of rolls for Mr. Slick's die.

more than the expected value of 0.17. Are these differences statistically significant? A chi-squared goodness-of-fit test will tell me.

Under the null hypothesis of a fair die, the expected frequency for each cell, or number, is $E_i = 60 \times 1/6 = 26.67$, plenty big enough to satisfy the sampling requirements of the test. The chi-squared statistic for this test, as calculated from these E_i and the observed frequencies listed in Figure 4.9 is

$$X^2 = \frac{(34 - 26.67)^2}{26.67} + \frac{(39 - 26.67)^2}{26.67} + \frac{(22 - 26.67)^2}{26.67} + \frac{(26 - 26.67)^2}{26.67}$$
$$+ \frac{(28 - 26.67)^2}{26.67} + \frac{(11 - 26.67)^2}{26.67} = 17.8.$$

The $1 - \alpha = 0.95$ critical value for the chi-squared distribution with $M - 1 = 5$ degrees of freedom is $\chi^2_{0.95,5} = 11.1$. Since the test statistic exceeds this critical value, I must conclude my test subject was playing with a loaded die. And Mr. Slick is, in fact, a cheater.

The chi-squared test is a versatile technique that can be used to answer many common questions about discrete data. But this test isn't only for dice games and surveys. There are other variations of this test, variations that can be used to test continuous data. These will be presented in following chapters.

BIBLIOGRAPHY

Corder GW, Foreman DI. *Nonparametric Statistics for Non-Statisticians: A Step-by-Step Approach*. Hoboken: John Wiley & Sons, Inc; 2009.

Frenz R. Types of Likert Scales. Available at http://www.ehow.com/list_6855863_types-likert-scales.html. Accessed May 14, 2012.

GoingtoVegas.com. Learn to Play Craps Guide. Available at http://goingtovegas.com/kpv-crap.htm. Accessed July 13, 2013.

Hollander M, Wolfe DA. *Nonparametric Statistical Methods*. New York: John Wiley & Sons, Inc; 1999.

Jarman K. *The Art of Data Analysis: How to Answer Almost Any Question Using Basic Statistics*. Hoboken: John Wiley & sons, Inc; 2013.

Triola MF. *Elementary Statistics*. 11th ed. Toronto: Pearson; 2011.

How to Play Sixes Bet. Stormdark I.P. and Media. Available at http://www.dice-play.com/Games/SixesBet.htm. Accessed May 12, 2014.

5

WHY IT PAYS TO BE A STABLE MASTER: SUMO WRESTLERS AND OTHER ROBUST STATISTICS

Sumo wrestling is the national sport of Japan. Two large men, dressed in loincloths, with long hair tied neatly into sumo knots, charge at one another like raging bulls. The first one to knock his opponent down or push him outside the tiny wrestling ring wins the match. These wrestlers are robust by any definition: disciplined, dedicated, and of a seriously hearty constitution. They live in *heya*, or stables, compounds separated from the rest of society and managed by a stable master. They train almost constantly. Their lives are highly structured, with everything from their sleeping schedule to their clothing dictated by tradition. And they eat a diet carefully designed to help them gain as much weight as possible. The good ones earn a nice living, have a loyal fan base, and live a life of relative comfort. The bad ones earn almost nothing while they spend their lives serving the better wrestlers.

Like baseball, football, basketball, and well, most other sports, Sumo wrestling has had its share of scandals. In 2007, a young sumo wrestler died in a bullying incident involving a several wrestlers, a stable master, and a beating with a large beer bottle. Two years later, a prominent sumo wrestler and his stable master were caught betting illegally on, of all sports, baseball. In 2011, a number of wrestlers and stable masters were caught rigging matches for money. This last scandal caused so much concern, the Japanese Sumo Organization cancelled the national tournament in November of that year.

Beyond Basic Statistics: Tips, Tricks, and Techniques Every Data Analyst Should Know,
First Edition. Kristin H. Jarman.
© 2015 John Wiley & Sons, Inc. Published 2015 by John Wiley & Sons, Inc.

Suppose I'm a Japanese Sumo stable master involved in the recent scandals. Four of my top wrestlers have been accused of match rigging. The others are disgruntled. My entire stable has been suspended for three years, until I can prove I've cleaned up my act. Three years is a long time to go without any winnings, and if I don't do something, I'll never be able to afford the rather hefty costs of feeding and housing my men until the suspension expires. In other words, my days as a stable master are over.

Or are they?

Japan isn't the only country that loves to watch a contest between men who've trained beyond their physical limits. Across the Pacific Ocean is a country full of people that spend their weekends attending, watching, betting, and cheering on such competitions. What if I take my show on the road? To the land of opportunity. To America.

Americans are no strangers to sports scandals, so my reputation for match rigging shouldn't bother them. And the United States is the great spectator nation. If these people are willing to pay money to attend a medieval-themed restaurant and eat turkey legs while watching a fake jousting match, then surely they'll be willing to pay to eat teriyaki chicken on a stick while watching two half-naked 600 pound men slamming into each other. The opportunity is ripe. I just need to pick a handful of my heartiest wrestlers and arrange exhibition matches in all the major American cities. And to make sure this tour is a success, I'll rely on good marketing, great wrestling, and some carefully chosen robust statistics.

DESCRIPTIVE STATISTICS: A REVIEW
FOR THE *JONOKUCHI*

Japanese Sumo wrestlers are ranked according to a highly structured pyramid system. The *Yokozuna* lie at the top of the pyramid. These men are the most experienced, most successful wrestlers. The *Jonokuchi* lie at the bottom of the pyramid. *Jonokuchi* wrestlers are typically new to the sport. They haven't got many professional matches under their belt, and they're spending a lot of time developing basic skills. This section is for the statistical *jonokuchi* out there, those of you who are new to data analysis or in need of a tutorial on basic descriptive statistics. If this isn't you, feel free to skip to the next section.

Chapter 4 presented techniques for analyzing qualitative data, those observations that fall into descriptive categories or take on a few distinct numerical values. Methods for analyzing qualitative data tend to focus on frequencies—probabilities, percentages, and proportions for outcomes of interest. This allows a data analyst to turn categorical observations, such as hair color, into numerical data, such as the proportion of a sample having blond hair, so that statistical methods can be used to analyze it. This chapter and those that follow are devoted to quantitative data, numerical observations that have a meaningful ordering to them. Because quantitative data are numerically meaningful, the full force of mathematics can be put to work on it.

There are various techniques for analyzing quantitative data, and most of them rely on descriptive statistics. A descriptive statistic is a value calculated from a sample that estimates some property of the population under study. For quantitative data, the most common descriptive statistics are the sample mean, or average, and the sample standard deviation. The sample mean measures the center location of a sample. The sample standard deviation measures the variation around the sample mean. Together, these two descriptive statistics are used as a foundation for confidence intervals, hypothesis tests, regression analysis, and just about any other statistical technique you can name.

Three Things You Should Know About the Sample Mean

The sample mean is the average. It's calculated as the sum of all the observations, divided by the total number of observations. For example, the dataset 6, 7, and 8 has a sample mean $(6 + 7 + 8)/3 = 7$. This value describes the arithmetic center of a dataset, or if you imagine your data as a cloud of numbers on a number line, the middle of this cloud. The sample mean is one of the most common, and the most powerful, ways to measure the center of a population, but it isn't perfect. No statistic is. Here are a few things you should know about this popular statistic.

The Sample Mean Is a Good Estimate of the Population Mean   In a typical population, not all members are identical to one another. In other words, there's variation. To deal with this variation, statisticians have invented the probability distribution, a mathematical function that describes the behavior of a population. This function can be used for many things, for example, to calculate the critical value for a hypothesis test, or to predict the proportion of the population that will fall in some range of values. There are many common probability distributions, for example, the normal distribution and the Student *t*-distribution, and most of them are characterized by their mean and variance. However, in the real world, the data analyst almost never knows the population mean and variance ahead of time. These two parameters must be estimated. The sample mean is one way to estimate the population mean.

Just how good is it? If you were to grab a random sample from a population and calculate the sample mean, you'd get some (hopefully accurate) estimate of the population mean. If you were to do this again, you'd get another (hopefully accurate) estimate of the population mean. You could do this over and over again, each time getting a slightly different estimate. This collection of estimates would have some underlying probability distribution associated with them, complete with a mean and a variance. To determine if the sample mean is a good estimate, statisticians look at two important statistical properties of this estimate: bias and consistency.

Bias measures the difference between the expected value of an estimate and the corresponding parameter for the population. If the difference is zero, meaning the expected value of the estimate is the same as the population parameter, then we say it's an **unbiased estimate**. The sample mean is an unbiased estimate of the population mean. In other words, if you grab a random sample and take the average, you can expect this average value to be the population mean, plus or minus some error caused by using a subset of the population to estimate the true value.

Just how big is the error in the sample mean? The accuracy of this estimate is typically measured using the standard error. In general, the **standard error** is the variance of an estimate. This value tells you how much you can expect your estimate to deviate from the true, underlying population value. If you have a sample size of N, and if σ is the variance of the original observations in the sample, then the standard error of the sample mean is $\sigma / \sqrt{N}$. What does this mean for a typical data analysis? Among other things, it means that you can make the sample mean as accurate as you'd like, simply by adding more samples to your dataset. If you have nine samples, each with a standard deviation of one, the variance of the sample mean is 1/3. If you increase the sample size to one hundred samples, the variance of the sample mean drops to 1/10.

An estimate is **consistent** if, as the sample size grows, the estimate grows progressively closer to the true value. Because the variance in the sample mean decreases every time N is increased, the sample mean is a consistent estimate of the population mean.

Two Very Different Datasets Can Have the Same Sample Mean   The sample mean is a very useful statistic, but any time you compress many observations down into a single estimate, information gets lost. Two datasets may look completely different and yet have the same sample mean. Two datasets may look very similar and have different sample means. This is illustrated in Figure 5.1. The first dataset consists of discrete values between four and sixteen, with a sample mean of ten. The second dataset also has a sample mean of ten, but these values are continuous numbers, more tightly clustered around the center value. The third dataset looks similar to the second, with a similar range of continuous values clustered around the center value, but in this case, the sample mean is 11.6.

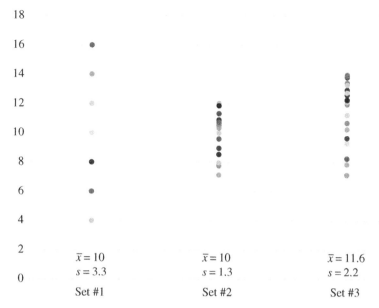

FIGURE 5.1   Three different samples, two different sample means.

What does this mean when it comes to data analysis? In a nutshell, the sample mean is a good measure of the center location of a population, but by itself, it only tells part of the story. Other descriptive statistics are also important when characterizing a dataset.

Extreme Values and Skewed Data Can Impact the Sample Mean   The United States is a big place. As I start planning what I've dubbed the Japanese Sumo Invasion Tour, I discover that America has many stadiums and arenas for rent, and they range in seating capacity from a few hundred to tens of thousands. Booking the right venue is important, too big and I'll end up paying for unsold seats, too small and I'll end up turning away paying customers with a sold out show. How many sumo fans can I expect to attend these events? To find out, I turn to USA Sumo, an organization that arranges amateur sumo tournaments around the country. This organization posts useful tidbits of information on its website, including the weight and record of competitors in the tournament as well as attendance levels at events over the past decade. Figure 5.2 shows a plot of 2013 USA Sumo attendance data (y-axis) for all cities (x-axis), listed in alphabetical order.

When we picture the sample mean of a dataset, we generally picture a value in the middle of the data, where the majority of observations lie, with about half the observations falling below the mean and half falling above it.

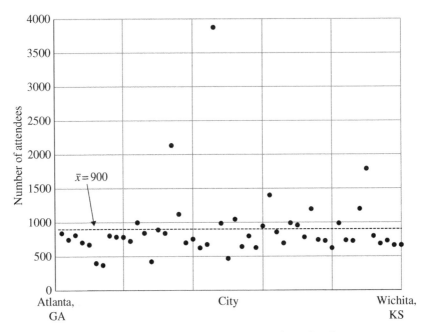

FIGURE 5.2   USA sumo games attendance by city.

The sample mean of the attendance data is $\bar{x} = 900$. The vast majority, 75%, of the values fall below this mean. Why is the sample mean of the attendance data so high?

Two reasons: outliers and skewed data.

Outliers are extreme data values, observations that sit by themselves, far away from the center of the data cloud. There are three obvious outliers in Figure 5.2, the three events with more than 1500 attendees. Outliers tend to shift the sample mean toward the extreme values. In this case, all three of the outliers are large values, and so they increase the sample mean.

Skewed data can be best illustrated using a histogram. A **histogram** is a frequency distribution of continuous data. It's just like a frequency distribution for discrete data, except for one thing. Because the data are continuous, the observations are first dropped into equally spaced bins spanning the range of values so that a bar chart of frequencies can be constructed. Figure 5.3 illustrates a typical histogram. The observations are symmetrically clustered around an obvious center point, and most of the observations lie close to this value. Away from center, the number of observations decreases smoothly to zero. A histogram with this shape is called a bell-shaped histogram. Many types of data naturally conform to this shape, and when they do, the sample mean falls right in the middle of the distribution where the top of the bell lies.

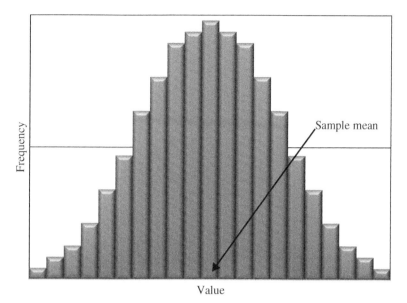

FIGURE 5.3   The bell-shaped distribution.

Figure 5.4 shows the histogram of the 2013 USA Sumo attendance data. These data do not follow a bell-shaped distribution. There are fewer observations below the highest frequency bin than above it. In other words, it looks like a bell-shaped distribution that's been stretched to the right. In statistics, we say these data are **right-skewed**. Skewed data will tend to shift the sample mean in the direction of the stretch, in this case to the right. The result is a larger sample mean than you'd expect for the standard bell-shaped distribution.

Three Things You Should Know about the Standard Deviation

The **standard deviation** is the average deviation, or variation, of all the values around the center location. If your measurement values are x_1, x_2, x_3, ... x_N, and your sample mean is $\bar{x}$, the formula for the standard deviation is

$$s = \sqrt{\frac{(x_1 - \bar{x})^2 + (x_2 - \bar{x})^2 + (x_3 - \bar{x})^2 + \cdots + (x_N - \bar{x})^2}{N - 1}}.$$

Thinking of a dataset as a cloud of numbers along a number line, the standard deviation reflects the width of this cloud. The bigger the cloud, the bigger the standard deviation.

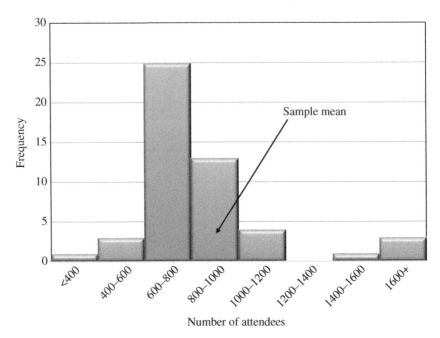

FIGURE 5.4   Histogram of 2013 USA sumo attendance data.

The Standard Deviation Is a Good Estimate of the Population Standard Deviation   The sample standard deviation can be evaluated just like the sample mean, in terms of its bias and consistency. The standard deviation is both unbiased and consistent. In other words, if you grab a random sample from your population and calculate the standard deviation, you can expect it to be pretty close to the population standard deviation, plus or minus some uncertainty caused by using a subset of the population to estimate it. And as you increase your sample size, you can expect the sample standard deviation to get closer and closer to the population standard deviation.

Extreme Values Can Impact the Standard Deviation   By itself, the sample mean provides some information about a dataset. Adding the standard deviation adds more. However, the standard deviation is even more sensitive to skewed data and outliers than the sample mean. For example, the sample standard deviation of the 2013 USA Sumo attendance data is $s=528$. Removing just three values, the three obvious outliers, from this dataset reduces the standard deviation to $s=203$. In other words, by removing just three extreme values out of a total of fifty, the standard deviation is cut to less than half its original value.

The Three-Sigma Rule Can Help You Understand Your Data   If a dataset has a nicely rounded, bell-shaped distribution like the one in Figure 5.3, over

99% of the observations will fall within three standard deviations of the sample mean. This rule can help you gain a little insight into your dataset. For example, the sample mean of the 2013 US Sumo attendance data is $\bar{x} = 900$, and the standard deviation is $s = 527$. According to the three-sigma rule of thumb, if these data are roughly bell-shaped, then over 99% of the observations should fall within three standard deviations of the sample mean. This means none of the fifty observations are expected to fall outside the range $(\bar{x} - 3s, \bar{x} + 3s)$, or $(-681, 2481)$.

The lower limit, a negative attendance value, is physically impossible. The smallest attendance occurred in Boise, Idaho, with 372 people, almost 900 more than this limit. When a three-sigma rule limit, particularly the lower limit, is far below the lowest meaningful value, this can be an indication of skewed data.

There is a single observation above the upper three-sigma rule limit of 2481, Los Angeles, with an attendance value of 3875. That's a difference of almost two standard deviations. When observations outside the three-sigma range are far from the upper or lower limits like this, they're obvious outliers that could very well be impacting the sample mean and standard deviation. More information on this and other outlier detection methods is provided in Chapter 7.

THE JAPANESE SUMO INVASION: WHY IT PAYS TO BE ROBUST

Figure 5.5 shows the 2013 USA Sumo attendance data, this time with the four most attended cities labeled. From the plot, it's easy to see at least three of these four shows were outliers. While the average attendance at the grand sumo tour events was 900, the Los Angeles event saw a whopping 3875 spectators. Honolulu and San Francisco were second and third with 2134 and 1791 attendees, respectively. New York was fourth with 1491 attendees.

These four observations clearly influence both the sample mean and standard deviation. The sample mean with these four values included is 900, but when they're left out of the calculations, it drops to 776. The standard deviation with these four values is 527, and when they're excluded, it drops to 170. But what, if anything, should be done about these extreme observations? Are they truly different from the rest of the population, needing different treatment? Or are they perfectly legitimate values that just happen to be big? If I simply remove them from the dataset, then the sample mean and standard deviation will certainly drop, but will they more accurately represent the underlying population?

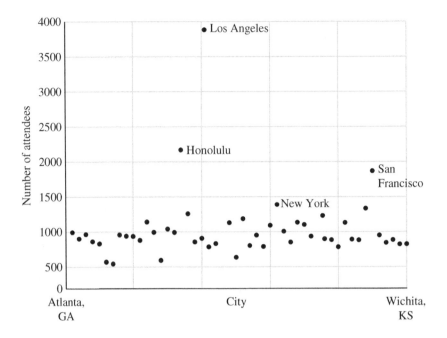

FIGURE 5.5   USA sumo games attendance with outliers.

These are questions every data analyst must answer when confronted with outliers, and these are the questions I face as I think about booking venues for my Japanese Sumo Invasion Tour. There are formal strategies for identifying and eliminating outliers, and I could use the methods from Chapter 7 to do this. But time is money and I don't have much of either. So rather than wading through the data, trying to determine which values to keep and which to throw out, I'm going to estimate the sample mean and standard deviation using a different approach, one that automatically takes care of the outliers for me.

There are two types of estimates that can help me with this problem. **Robust estimates** work well when your data do not follow a bell-shaped distribution. **Resistant estimates** aren't influenced by outliers. Robust and resistant statistics are like two sumo wrestlers from the same stable. They both have similar characteristics, and even though the stable master knows they're completely different, a casual observer may have trouble telling them apart. Robust and resistant estimates are evaluated differently—one in terms of robustness to non-normal distributions and the other in terms of resistance to outliers—but the end result is typically the same. When an estimate is robust, it's usually resistant, and when it's resistant, it's usually robust. Therefore, you don't need to be a stable master, highly trained in the subtleties of robust and resistant statistics, to be able to use them.

Summarizing a Sample with Percentiles

Percentiles are descriptive statistics taught in most introductory courses. Considered to be both robust and resistant, a **percentile** is the value below which some specified percent of the data fall. It's convenient to think of a percentile in terms of a sorted list of data values. Specifically, if you sorted your observations from smallest to largest, the 20th percentile would be the observation 20% of the way down this list. For a sample with N observations, this would be the $0.20 \times Nth$ value, rounded up to the next whole number as needed. For example, for the ten sample sorted data 1, 3, 3, 4, 5, 5, 5, 6, 6, 9, the 20th percentile is the $0.2 \times 10 = 2$nd value down the list, which is 3. The 75th percentile is $0.75 \times 10 = 7.5$, rounded up to the eighth value, which is 6.

A percentile can be anything, 10th, 30th, or 75th, but some percentiles are more meaningful than others. The 50th percentile, the observation halfway down the sorted list of values, is the **median**. The median is a robust alternative to the sample mean for describing the center location of a dataset. Together, the 25th, 50th, and 75th percentiles are called the **quartiles** of a dataset. The 75th percentile and 25th percentile, also called the third and first quartiles, can be used to measure variation. Specifically, the **interquartile range**(IQR) is the 75% percentile minus the 25% percentile. For the following dataset—1, 3, 3, 4, 5, 5, 5, 6, 6, 9—the median is five and the interquartile range is $6-3=3$.

The quartiles together with two more descriptive statistics, the minimum and maximum, make up what's called a **five-number summary**. A five-number summary is a quick and powerful way to gain a little insight into a dataset. For example, Figure 5.6 shows a five-number summary of the 2013 USA Sumo attendance data. The attendance numbers span from 372 to 3875. That's a total range of $3875-372=3503$. The interquartile range, $953-689=264$, is just a small fraction of this overall range. The 75th percentile is below 1000, less than a third of the maximum value.

This quick numerical comparison can be helpful, but I always find a graph to be more informative. A **boxplot** is the graphical representation of a five number summary. Figure 5.7 shows a boxplot of the sumo attendance data summarized in Figure 5.6. There's a box with a line through it. The line through the center of the box represents the median—the 50th percentile. The top and bottom edges of the box represent the 25th and 75th percentiles, respectively. The height of the box is the IQR. Whiskers stretch above and below the box to the extreme values, in other words, the minimum and maximum.

In general, a nice symmetric, bell-shaped dataset will have a box with a line traveling right through the center of it. Each of the whiskers will be a little longer than the box is tall. Figure 5.7 looks nothing like what you'd

Minimum	372
25th percentile (1st quartile)	689
50th percentile (median, 2nd quartile)	783
75th percentile (3rd quartile)	953
Maximum	3875

FIGURE 5.6   Five number summary of the USA sumo attendance data.

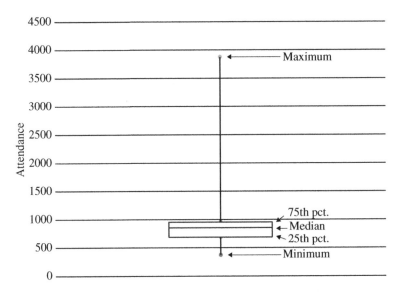

FIGURE 5.7   Boxplot of 2013 US sumo attendance data.

expect from normally distributed data. The median line travels roughly through center of the box, and the lower whisker appears to be just a bit longer that the height of the box. But the upper whisker dominates this plot. It's much longer than it should be for normally distributed data. This suggests at least one outlier.

Robust Center Location

For approximately normal data with no outliers, the sample mean is the *Yokozuna*, the grand champion, the best of the best, when it comes to estimating the center location of a dataset. It's unbiased, it's consistent, and results like the central limit theorem make it easy to work with in a typical basic statistical analysis. However, when it comes to robust and resistant estimates for center location, no clear winner exists. Over the years, several techniques have been developed, and each technique has its strengths and weaknesses.

There are several criteria for evaluating robust or resistant statistics. One I find particularly useful is the breakdown point. The **breakdown point** is a method for evaluating resistance to outliers. It's the proportion of extreme values a statistic can handle before it starts to become impacted by these values. Conceptually, you can find the breakdown point of an estimate by taking observations one by one and pushing them to the extreme, making them as big (or small) as you like. The largest proportion of observations you can change without changing the estimate is the breakdown point.

For example, take the dataset 1, 3, 3, 4, 5, 5, 5, 6, 6, 9. The sample mean is 4.7. If I take the most extreme value, 9, and change it to some really big number, say 1000, this change increases the sample mean to 103.8. Since turning only a single value into an outlier alters the sample mean, no changes can be made without changing the estimate. The breakdown point of the sample mean is zero. It may be the heavyweight champion of all center location estimates, but it's easily toppled by outliers.

Median   The median is a robust and resistant alternative to the sample mean. If your data cloud has a typical bell-shaped frequency distribution, the median will fall right in the center, close to the sample mean. If your data cloud is not typical or has extreme values, then the median can be quite different than the sample mean. For example, the 2013 USA Sumo attendance data have a sample mean of 900, and a median of 783. That's a difference of over 100 attendees. For my USA Sumo Invasion Tour, that translates into more than one hundred tickets per show, and more than 100 customers buying merchandise.

Where the sample mean has the lowest possible breakdown point, the median has the highest. Consider the dataset 1, 3, 3, 4, 5, 5, 5, 6, 6, 9. The median of these data is 5. Increasing the most extreme value, 9, to 1000, doesn't change the median. It's still 5. Increasing the next highest value, 6, to 1000, doesn't change the median. It remains 5. In fact, you can keep doing this until up to half of the values have been pushed to the extreme, and the median remains the same. In other words, the breakdown point of the median is 50%.

Aside from its resistance to outliers, the median has another advantage. Where confidence intervals for robust and resistant estimates can be difficult to find, calculating a confidence interval for the median involves little more than the binomial distribution. Here's the process. If you recall, the median of a population is the midpoint, the 50th percentile. If you pick a single observation from the population at random, the probability it will be greater than the median is one-half. If you sample N observations, the number n that are greater than the median is a binomial random variable, in other words, it follows a binomial distribution with N trials and success probability $p=1/2$. The probability n is within some range, n_L to n_U, is $P\{n_L<n<n_U\}$. And for this probability to be $1-\alpha$, in other words, for a $1-\alpha$ confidence interval, you can choose an upper limit n_U for which $P\{n>n_U\}=1-\alpha/2$ and a lower limit n_L for which the probability $P\{n<n_L\}=\alpha/2$.

The values n_L and n_U are not the confidence limits on the median of a sample, rather, they're the lower and upper confidence limits *on the number of observations in a sample size of N that exceed the median*. To convert these values into the confidence limits for the median of a sample, you sort your observations from smallest to largest. The n_L th smallest value is the lower limit on this confidence interval and the n_U th smallest value is the upper limit.

A confidence interval for the median USA Sumo attendance can be calculated in this way. There are $N=50$ events and the median attendance is 783. For a 95% confidence interval, the $\alpha/2=0.025$ critical value for the binomial distribution with $N=50$ and $p=0.5$ is $n_{lower}=18$. The $1-\alpha/2=0.975$ critical value for this same distribution is $n_{upper}=32$. Therefore, the lower bound is the 18th smallest value, which happens to be 725, and the upper bound is the 32nd smallest value, which is 814. So the 95% confidence interval for the median attendance is (725, 814).

Trimmed Mean   If the median has such a great breakdown point, why would anybody ever need another robust or resistant measure of center location? It turns out, the breakdown point isn't the only useful way to evaluate a robust or resistant statistic. Relative efficiency is also important. The **relative efficiency** measures the variance of a statistic relative to the sample mean, specifically,

$$\text{Efficiency of } E = \frac{\text{Variance of the sample mean}}{\text{Variance of } E}$$

The variance of the sample mean, the standard error, is $\sigma/\sqrt{N}$. Because the sample mean uses all N observations in its calculation, the variance of this estimate is as low as possible. Robust and resistance estimates tend to ignore some observations, and this gives them a larger variance, or uncertainty, than the sample mean. In other words, this gives them an efficiency of less than one. For example, because the median value is calculated from one or two observations, it's variance is larger than the standard error. In fact, for large, well-behaved samples, the median only achieves a maximum efficiency of 0.64, or 64%.

When it comes to estimating central location, the trimmed mean is more efficient than the median. The **trimmed mean** is calculated by trimming, removing a specified number of high and low values, and calculating the average of the remaining values. For example, the 20% trimmed mean is calculated by removing 20% of the values—the largest 10% and the smallest 10%—and then calculating the average from what's left. To calculate the 20% trimmed mean of a dataset with ten values, you'd remove the smallest value and the largest value and calculate the average of the remaining values. For example, with sorted data 1, 3, 3, 4, 5, 5, 5, 6, 6, 9, you'd remove the 1 and the 9, and calculate the average $(3+3+4+5+5+5+6+6)/8=4.6$.

The trimmed mean offers a higher resistance to outliers than the sample mean, and better efficiency than the median. The X% trimmed mean has a breakdown point of $X/100$, meaning up to X% of the observations can be outliers without impacting the estimate. For example, in the previous list of data values, you could decrease the lowest value as much as you want and increase the largest value as much as you want and still get a 20% trimmed mean of 4.6. And because it uses $100-X$% of the observations, the trimmed mean has an efficiency of $1-X/100$. For a 20% trimmed mean, this gives an efficiency of 0.8, or 80%.

How do the sample mean, median, and trimmed mean compare when confronted with the Sumo attendance data? Figure 5.8 shows the results. The sample mean is higher than the median, due to the presence of a few very large values. The trimmed mean lies somewhere between the sample mean

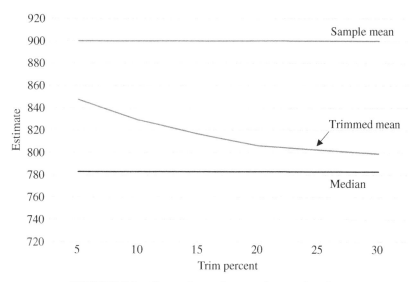

FIGURE 5.8 Comparing estimates of center location.

and median. When the trimmed percent is low, the trimmed mean uses more of the observations and looks more like the mean. As this percent increases, the trimmed mean uses fewer and fewer observations and starts to look more like the median.

Robust and Resistant Variation

Like the sample mean, the sample standard deviation uses all of the observations in a sample. And so, like the sample mean, this statistic has a breakdown point of zero. In other words, the standard deviation is just as, if not more, sensitive to outliers than the sample mean. There are two commonly used robust and resistant alternatives to the standard deviation, both of them based on percentiles.

The Interquartile Range   The IQR has already been introduced as a way to measure variation. Also called the middle fifty, this statistic measures the range of the middle 50% of observations in a sample. The IQR has a breakdown point of 25% meaning it can withstand up to 25% of the data being outliers. For a symmetric, bell-shaped dataset, the IQR should be just a little larger than the standard deviation, about 1.35 times larger. For the USA Sumo attendance data, however, this value is only about half the standard deviation, with IQR $= 264$ and $s = 527$. This illustrates just how much a few outliers can affect the standard deviation.

The Median Absolute Deviation   The **median absolute deviation** relies heavily on the median to measure variation. If m is the median of your data, the median absolute deviation (MAD) is

$$\text{MAD} = \text{median}(|x_1 - m|, |x_2 - m|, \cdots, |x_N - m|)$$

In other words, it's the median of the absolute differences between every data value and the median of the data values. For example, the dataset 1, 3, 3, 4, 5, 5, 5, 6, 6, 9 has a median of 4.5. The MAD is the median of the values: $|1-4.5| = 3.5$, $|3-4.5| = 1.5$, $|3-4.5| = 1.5$, $|4-4.5| = 0.5$, $|5-4.5| = 0.5$, $|5-4.5| = 0.5$, $|5-4.5| = 0.5$, $|6-4.5| = 1.5$, $|6-4.5| = 1.5$, and $|9-4.5| = 3.5$. And this median turns out to be 1.5.

Because it relies on the median, which has a breakdown point of 50%, the median absolute deviation also has a breakdown point of 50%. This value tends to be smaller than the standard deviation. For a normal distribution, the MAD is about 67% of the standard deviation. Not so for the USA Sumo attendance data. For this dataset, the standard deviation is five times the MAD value of 109, giving further evidence of outliers in the data.

Robust Confidence Intervals

A good estimate should have a margin of error associated with it, in other words, a confidence interval. Estimating the confidence interval of a sample mean is straightforward. There are simple formulas for this and most data analysis software packages make them readily available. Confidence intervals for resistant or robust statistics aren't always so straightforward. The formulas for these types of intervals typically require advanced statistics and some rather hefty approximations, and they're often only applicable under very special circumstances.

Rather than relying on complicated and specialized formulas to calculate the confidence interval of a robust or resistant estimate, many data analysts turn to a widely used technique: bootstrapping. **Bootstrapping** is a data-based method for calculating the variance, bias, or a confidence interval of an estimate. Bootstrapping doesn't require complicated formulas or approximations. You only need a way to randomly select observations from your dataset using a technique known as resampling.

Resampling is just like sampling, except now you're treating your dataset as the entire population. In resampling, you take your original sample of N observations and randomly choose N observations, replacing each one after you pick it. In this way, you'll get a new sample of the same size as the original sample where some observations might have been chosen twice or three times, and others none at all. From this new (re-)sample, you can calculate whatever estimate you'd like, say the median or the trimmed mean. If you do this—resample and estimate—many times, you get a large number of estimates, and you can use these estimates to calculate variance, bias, or confidence intervals.

The process of resample bootstrapping goes like this:

1. Generate a new sample from your original dataset by resampling N observations.
2. Calculate your estimate of choice using this resampled data.
3. Repeat steps 1–2 many times (500+, the more the better), saving the value of the estimate each time.
4. Use the estimates from the resampled data to calculate your chosen measure of accuracy (bias, variance, confidence intervals).

For example, suppose I wanted to use resample bootstrapping to calculate a 95% confidence interval for the median of the attendance data. I'd take this original dataset of $N = 50$ values and resample to get a new set of $N = 50$ values. I'd calculate the median of this resampled data and save it.

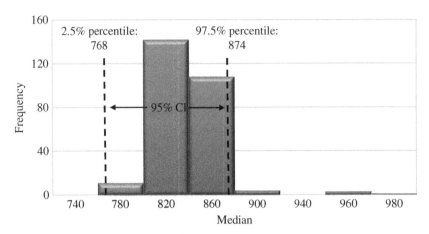

FIGURE 5.9   Bootstrap estimate of standard deviation and 95% confidence interval of median.

I'd repeat this process many more times until I had, say, 500 median estimates from as many different resampled datasets. I'd then find the 95% confidence interval by calculating the 2.5 and 97.5% percentiles from these median values. These percentiles would be the lower and upper limits on my confidence interval. Figure 5.9 shows the results of this bootstrapping process.

WHEN ROBUST DOES IT BETTER

With all these different ways to estimate center location and variation, you might be wondering when you should stick to the sample mean and when you should look to the median. When you should use the IQR and when you should go with the MAD. Unfortunately, the answer to this question is just as much opinion as it is statistics. There are some statisticians who rarely, if ever, use robust or resistant statistics, preferring instead the classic sample mean and variance. Others argue that with good robust and resistant statistics like the trimmed mean and MAD, there's no reason ever to go back to the traditional mean and standard deviation. Here are some things to consider as you form your own opinion on this matter.

During the initial exploration of a dataset, especially when this dataset is large, it's often useful to calculate both resistant statistics and traditional statistics. Comparing them can help you understand your data better. It can point to outliers and skewed distributions, suggest excessive variation, or confirm that you have a well-behaved, bell-shaped dataset.

There are many, many statistical techniques that rely on the traditional sample mean and standard deviation, and these are the ones most commonly found in a typical data analysis software. In other words, the traditional statistics are the easiest to use. So, many data analysts prefer to inspect the data for outliers and remove errors or faulty observations (see Chapter 7). This not only helps you verify the observations are legitimate but also allows you to use the tried and true statistics and the tried and true methods that go along with them.

In some cases, you just can't inspect each and every data value for outliers. Examples of situations where it's difficult or impossible to do this include real-time monitoring of stock market data, quality control situations where new observations are constantly arriving, or big data applications where you have so many observations, your computer chokes whenever you try to load it all into memory. In cases like these, it makes sense to turn to resistant or robust measures of center location and variation in order to prevent unusual data from ruining an otherwise perfectly good statistical analysis.

HARVESTING THE AMERICAN DREAM

As a Japanese stable master, I don't know much about Americans. But I do know that they love big things. SUVs. Super-sized combo meals. My stable of sumo wrestlers should fit nicely into this bigger-is-better model. The question is, are they big enough?

In recent years, sumo wrestling has been slowly migrating out of Japan, making its way into other countries. America has its own version of sumo wrestling. Like the country itself, American sumo wrestling is a melting pot of people from around the world. It allows amateurs—both men and women—to come together and compete, just for the joy of competing. No decades of tightly regimented training. No rigid hierarchy. No living in a stable, cut off from the rest of society. Just wrestlers coming together to fight and have fun. While this American-style sumo throws away many long held traditions, it can help me. Specifically, by looking at the competitors in these events, I can get a good idea of what Americans expect out of a sumo wrestler.

American sumo has weight-classes: lightweight, middleweight, heavyweight, and open. I'm not interested in the smaller men, those weekend warriors who wrestle for nothing more than a good thrill. I'm only concerned about the serious contenders, the biggest and strongest men. The heavyweights. Figure 5.10 shows a boxplot of the heavyweights, along with some pertinent statistics.

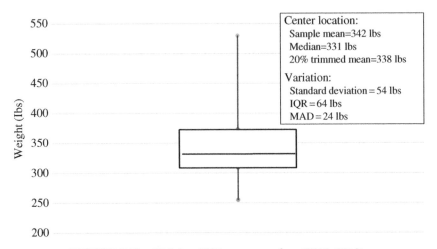

FIGURE 5.10   Weight of US sumo wrestlers (2000–2013).

A man has to weigh at least 253 lbs to wrestle in the heavyweight class, so I'd expect the distribution of these data to be right-skewed: with observations starting abruptly at 253 lbs, tapering off to the right as the weight increases. The boxplot shows this. For the $N = 125$ wrestlers in this study, the weight ranges from the lower limit of 253 to a whopping 530 lbs. The median line runs not quite through the center of the box, but just below it. The lower whisker is about the length of the box height, a little shorter than you'd expect for a normal distribution. The upper whisker is much longer.

What do the descriptive statistics show? First, the mean, median, and trimmed mean are within eleven pounds of one another, a small difference when it comes to 300 pound wrestlers. The standard deviation is $s = 54$. If the data are approximately normal, then the IQR should be about 1.35 times this value, or 74 lbs, and the MAD should be about 0.67 times this value, or 36 lbs. In reality, the IQR is sixty-four and the MAD is twenty-four. The distribution of these data is definitely non-normal.

The standard 95% confidence interval for the mean USA Sumo weight is (332, 352), however, because these data are non-normal, this estimate may not be the best choice. How does a robust estimate—the 20% trimmed mean, for example—compare to it? To find out, I've used bootstrapping to construct a confidence interval for this estimate. Specifically, I resampled the heavyweight data so that I had 500 samples, each containing $N = 125$ values. Figure 5.11 shows the bootstrap estimate of the 95% confidence interval for the 20% trimmed mean. This bootstrap confidence interval is (329, 348), just a few pounds shy of the traditional confidence interval based on the normal distribution.

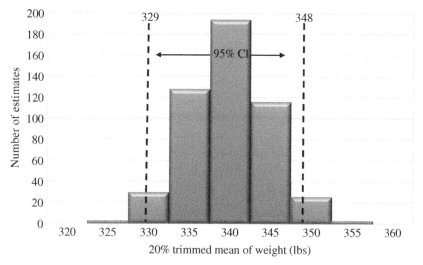

FIGURE 5.11 The confidence interval for the trimmed mean of USA sumo heavyweights.

So, what do Americans expect out of a sumo wrestler? According to my analysis, a typical heavyweight is somewhere between 329 and 348 lbs. The largest heavyweight is much heavier than this, though, tipping the scales at 530 lbs. My stable of wrestlers average 395 lbs, definitely heavier than the average US amateur, but if I really want to cash in on the American dream, I need to bring my biggest and strongest wrestlers, those that make a 500 pounder look like a weakling. Those that can demonstrate there's nothing like authentic Japanese sumo wrestling to impress a crowd of thrill-hungry spectators.

BIBLIOGRAPHY

CNN.com. Sumo 'Fixing' Scandal Rocks Japan. Available at http://edition. cnn.com/2011/SPORT/02/03/sumo.japan.fix.hanaregoma/index.html. Accessed February 4, 2011.

Hollander M, Wolfe DA. *Nonparametric Statistical Methods.* New York: John Wiley & Sons, Inc; 1999.

Huber P, Ronchetti E. *Robust Statistics.* Hoboken, NJ: John Wiley & Sons, Inc; 2009.

Mishima S. Introduction to Japanese Sumo. Available at http://gojapan.about.com/ cs/sumo/a/sumo.htm. Accessed June 21, 2014.

Mosteller F, Tukey JW. *Data Analysis and Regression: A Second Course in Statistics.* Reading: Addison-Wesley, Inc.; 1977.

Nihon Sumo Kyokai Official Grand Sumo Home Page. All About Sumo. Available at http://www.sumo.or.jp/en/ticket/index. Accessed June 24, 2014.

Sanchanta M, Narioka K. Japan grapples with new sumo-wrestling scandal. Wall Street J Available at http://online.wsj.com/news/articles/SB10001424052748704 25630457532037069587864. Accessed June 21, 2010.

The Japan Times. Sumo. Available at http://www.japantimes.co.jp/sports_category/sumo/. Accessed June 14, 2014.

Triola MF. *Elementary Statistics*. 11th ed. Toronto: Pearson; 2011.

USA Sumo. Available at http://www.usasumo.com/. Accessed June 15, 2014.

6

FIVE-HOUR MARRIAGES: CONTINUOUS DISTRIBUTIONS, TESTS FOR NORMALITY, AND JUICY HOLLYWOOD SCANDALS

Hollywood, California, a magical place where the rich and beautiful live. A place where success is measured in starring movie roles, and incomes are measured in millions. A place filled with decadence and dreams and bikinis and botox.

No wonder ordinary people are fascinated with Hollywood, especially those scandals that leak from every pore of the city's perfectly tanned complexion. One of the most popular scandals seems to be the hook-up/break-up story, where we watch as two celebrities get romantic, get married, and get divorced. These marriages never seem to last long. Singer Britney Spears' marriage to Jason Alexander lasted just two days. Celebrity Kim Kardashian's marriage to Kris Humphries lasted just seventy-two days. Actress Drew Barrymore has had two short-lived marriages, one lasting five months and the other lasting just thirty days. The success of a normal marriage is typically measured in years, but stories like these make us wonder if the success of a Hollywood marriage should be measured in weeks or even days.

Is it really true that celebrities just can't stay committed? Or is our national hook-up/break-up obsession causing us to think this way?

Hypothesis tests are one of the most powerful techniques in the data analyst's arsenal. The most commonly used hypothesis tests rely on the

Beyond Basic Statistics: Tips, Tricks, and Techniques Every Data Analyst Should Know,
First Edition. Kristin H. Jarman.
© 2015 John Wiley & Sons, Inc. Published 2015 by John Wiley & Sons, Inc.

assumption of normality, in other words, that the data follow a normal distribution. Thanks to some very important statistical theorems, this assumption works in many situations. However, if you have small sample sizes or if you want to answer a question that goes beyond a simple test for the mean or proportion of a population, blindly assuming normality can lead to inaccurate p-values and false conclusions.

Tests for normality and continuous goodness-of-fit tests are tools no data analyst should be without. These hypothesis tests help us determine whether or not a sample is consistent with the normal (or some other) distribution. They help us make the proper assumptions about our data so we can apply the most appropriate techniques and reach accurate conclusions. In this chapter, tests for normality will be used to answer the question: Are Hollywood marriages normal?

THE NORMAL DISTRIBUTION: THE MOST ORDINARY OF ALL PROBABILITY DISTRIBUTIONS

As a discipline, mathematics tends to be black and white. Two plus two always equals four. The area of a rectangle is always its length times its width. Pi is always 3.14. But the world is a messy place. There are exceptions to every rule, and we almost never know what's going to happen in advance. Variation and uncertainty plague every aspect of our lives, including our data. To deal with all this uncertainty, statisticians have invented the probability distribution.

The discrete probability distribution was reviewed in Chapter 4. It's a mathematical function describing how discrete data tend to behave. A **continuous probability distribution** does the same thing for continuous data, observations, or measurements that can take on any value in some range. A continuous probability distribution relies on a continuous random variable. As introduced in Chapter 4, a **continuous random variable** is a variable, usually denoted by X or Y, that represents some as-yet-undetermined outcome of a random experiment. The continuous probability distribution measures the probability a random variable will take on certain values. This is usually done using the **cumulative distribution function (cdf)**, or $P\{X \leq x\}$.

There are many continuous probability distribution out there, and each one has certain circumstances where it's the perfect choice. Without a doubt, the most common continuous probability distribution is the normal distribution. The normal distribution describes a random variable that can take on any real value, and its probability distribution function has a nice-symmetric, bell-shape. There are two parameters associated with the normal

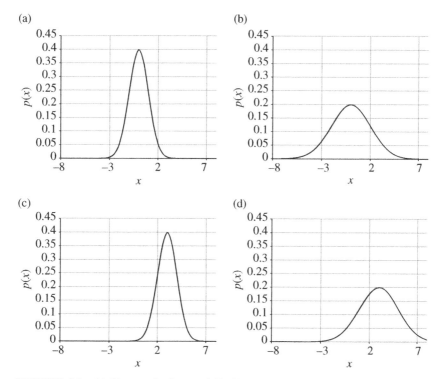

FIGURE 6.1 (a) The standard normal distribution (mean 0 and variance 1), (b) The normal distribution (mean 0 and variance 4), (c) The normal distribution (mean 3 and variance 1), and (d) The normal distribution (mean 3 and variance 4).

distribution: the mean and the variance. The mean determines the center location of the distribution, the point where the probability function is largest. The variance determines its width, whether it's tall and narrow or short and wide. Figure 6.1 plots the normal probability distribution function for different parameter values.

There's a good reason the normal distribution is so popular. Many types of data cluster symmetrically around some center value, and so many types of data naturally conform to the normal distribution. But that's only part of the reason this distribution is so widely used. If your data don't look at all normal, certain statistical techniques based on the normal distribution can still be used, thanks to a powerful result known as the central limit theorem. Confidence intervals for the mean, t-tests, F-tests, and analysis of variance are just some of the techniques that you can use, even if you don't have normal data. You only need a large enough sample. For more information on this topic, I refer you to any good basic statistics text. For my contribution, see Chapter 5 of *The Art of Data Analysis: How to Answer Almost Any Question Using Basic Statistics*.

As useful as the normal distribution is, sometimes a data analyst needs something more. If you have a small sample, for example, the central limit theorem no longer applies. When you want to predict future observations or estimate small probabilities, small deviations from the normal distribution can lead to big errors in the result. This is why every data analyst needs a test for normality in his toolkit.

NORMALITY TESTS EVERY DATA ANALYST SHOULD KNOW

Suppose I work as a scandal tracker for a major tabloid newspaper. My duties include dispatching reporters to all the Hollywood hot spots in search of juicy gossip. Movie stars are crafty and fickle, what's popular one week might be over and done the next. So, I want to analyze the locations of all the celebrity sightings I can find over the past few months. This includes stories in the tabloids as well as sightings posted on celebrity-watching websites such as smarp.com.

Most of the locations are old news to me. However, there's one spot, a parking lot in Santa Monica, CA, where a surprising number of celebrity sightings have occurred, at least a hundred in the past month. This parking lot is in front of a tidy, nondescript building that looks like it could be a doctor's office. There's no lettering on the outside of the building that might suggest what it is, and the location is unlisted. No phone number or business name available.

The celebrity sightings seem to come in waves, and there appear to be more sightings in the morning than any other time. To examine the distribution of these sighting times, I generate a histogram. I divide the day into one-hour bins, place every sighting into its appropriate time bin and plot the resulting frequencies. Figure 6.2 shows this histogram, with time of day plotted as number of hours since the earliest sighting, 8 A.M. Summary statistics and a normal probability distribution function with the same mean and variance have been added for reference.

The celebrity sighting dataset is a good example of right-skewed data. The sample mean is just under three hours after 8 A.M., at 10:45 A.M., but the mode, the maximum number of sightings occurs before that time, around 10 A.M. To the left of this maximum point, the frequencies drop off sharply (simply because there are no sightings before 8 A.M.). To the right of this maximum point, the frequencies look like a bell curve whose tail has been stretched. Many types of measurements tend to produce right-skewed data, among them are distance to some reference point, counts of people or items, and yes, the time until an event such as a celebrity sighting occurs.

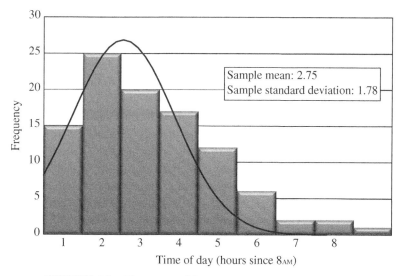

FIGURE 6.2   Histogram of Santa Monica celebrity sightings.

The data shown in Figure 6.2 could be used to construct a confidence interval for the mean. And with so many sightings ($N=100$), a standard interval based on normal data would be a perfectly legitimate choice. But the mean only measures the arithmetic center of a dataset, and I'm not really interested in the mean. If I send my best reporter to the parking lot to sniff around for a scandal, I'd like my chances of spotting a celebrity to be as high as possible. In other words, I'd like to send my reporter out when the probability of spotting a celebrity is at its highest. And according to the histogram, the mode, the bin with the highest frequency occurs not at the mean time of 10:45 A.M., but some time before that.

Since I'm not interested in a simple confidence interval or mean test, the distribution of my data matter. Figure 6.2 shows the data deviate from the perfectly normal bell shape. But how far off are they? Is this deviation statistically significant, or can I still apply tests based on the normal distribution? Here are some tools for answering these questions.

Q–Q Plots: A Picture Is Worth a Thousand Words

Just like the right photograph can reveal a celebrity scandal in ways words never could, the right graph can reveal hidden relationships, highlight outliers, and even help you determine if your data are normal. One of the best graphical tools for assessing normality is something called a Q–Q plot. This tool can be used to compare data to any discrete or continuous probability distribution. However, it's most often seen with the normal distribution.

The **Q–Q plot** is based on the concept of quantiles. **Quantiles** can be thought of as percentiles calculated at regularly spaced intervals. For a continuous random variable, the quantiles are calculated from the cdf or $P\{X \le x\}$ for specific values of x. Recall when x is at the lower end of the range of possible values, then $P\{X \le x\} = 0$. When x is at the upper end possible values, $P\{X \le x\} = 1$. Between those two points, the cdf gradually increases as x increases. If you divide the range 0–1 into some number of equally spaced intervals and calculate the x-values corresponding to the probabilities created by those intervals, you've got quantiles. For example, if you divide the probability range into three intervals you end up with two probabilities, 1/3 and 2/3. The x-values associated with these probabilities, where $P\{X \le x\} = 1/3$ and $P\{X \le x\} = 2/3$, are what we call 3 quantiles for this distribution. These are illustrated in Figure 6.3.

The number of intervals you choose is generally referred to by a number in front of the word *quantile*. Each successive interval is generally referred to by a number before that. In the example given earlier, with the interval divided into thirds, the quantiles are referred to as 3 quantiles. The first value, for which $P\{X \le x\} = 1/3$, would be the first 3 quantile. The second, for which $P\{X \le x\} = 2/3$, would be the second 3 quantile.

Quantiles can also be estimated from a set of data. To find quantiles of a data set, you sort the observations from smallest to largest, and work your way down

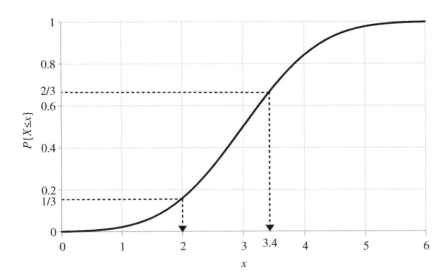

FIGURE 6.3 The 3 quantiles for a cumulative distribution function.

Sorted data values	2-quantiles	3-quantiles	5-quantiles	10-quantiles
1.1				1.1
1.2			1.2	1.2
1.3				1.3
1.4		1.4	1.4	1.4
1.6	1.6			1.6
1.7			1.7	1.7
1.7		1.7		1.7
1.8			1.8	1.8
1.9				1.9
2.0				

FIGURE 6.4   Quantiles for a dataset with $N = 10$.

this sorted list, flagging data values at regular intervals. For example, to find the 3 quantiles, you flag the values 1/3 and 2/3 of the way down your list. If an interval, such as 1/3, falls between two x-values, you simply round up to the next largest value. Figure 6.4 shows various quantiles for a small dataset.

If your data follow a normal distribution, then the quantiles should match the corresponding quantiles of the normal distribution. This is what a Q–Q plot shows. Specifically, a **Q–Q plot** is a plot of the data quantiles against the corresponding quantiles for a normal distribution. If the points fall close to the line $x = y$, your data are approximately normal. If your data significantly deviate from the line, then they aren't normal. Figure 6.5 illustrates a Q–Q plot.

Most statistical analysis software packages have a function for generating Q–Q plots. You only need to supply the data. If you don't have access to such software, it's easy enough to generate Q–Q plots in spreadsheet programs like Excel. Here are the instructions on how to construct a Q–Q plot in a typical spreadsheet program.

1. Sort the data values from smallest to largest. Place the sorted values in Column A.
2. Rank each data value by placing the number "1" in the first row, adjacent to the smallest value, the number "2" in the second row adjacent

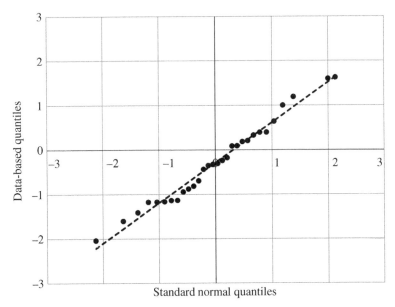

FIGURE 6.5   A Q–Q plot.

to the second smallest value, and so on up to a rank of N. Place the ranks in Column B.

3. Calculate cumulative probabilities for each data value as follows. In Column C, take the rank of each data value, subtract 0.5, and divide by N, for example, $C1 = (B1-0.5)/N$.

4. Calculate z-values for the normal distribution. In Column D, calculate the inverse normal function for the cumulative probabilities in Column C. For example, $D1 = \mathrm{norminv}(C1)$.

5. Calculate z-scores for the data. In cell F1, calculate the average of the data values in Column A. In cell F2, calculate the standard deviation of the data values in Column A. Back to Column E, calculate the z-scores of the original data by subtracting the sample mean and dividing by the standard deviation. For example, $E1 = (A1-F1)/F2$, $E2 = (A2-F1)/F1$, and so on.

6. Generate a scatterplot of column D versus column E. This is a Q–Q plot.

There are a variety of common patterns that appear in Q–Q plots, and these patterns suggest certain behaviors in data. For example, a Q–Q plot that resembles a tilted "U" suggests right-skewed data. A Q–Q plot that has an 'S' shape suggests light-tailed data, data with tails that are narrower than the corresponding bell curve. Figure 6.6 illustrates some common Q–Q plot patterns and shows what those patterns suggest about a dataset.

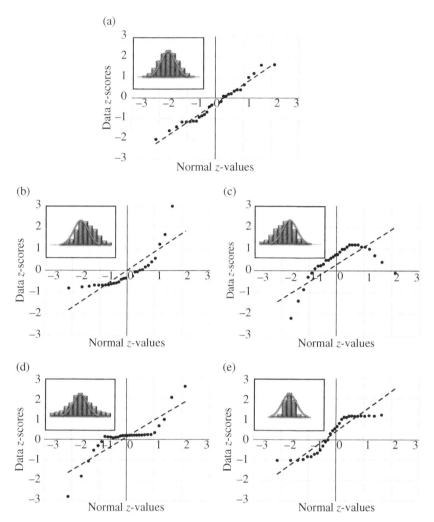

FIGURE 6.6 (a) Normal data, (b) Right-skewed data, (c) Left-skewed data, (d) Heavy-tailed data, and (e) Light-tailed data.

Figure 6.7 shows the Q–Q plot of the celebrity sightings data. Most of the values fall close to the line, but there is a noticeable "U" pattern in the values. This suggests the data are slightly right-skewed. This comes as no surprise since the histogram in Figure 6.2 suggested the same thing.

Hypothesis Tests for Normality

The celebrity sighting data are slightly skewed. Are they so skewed I need to abandon the assumption of normality?

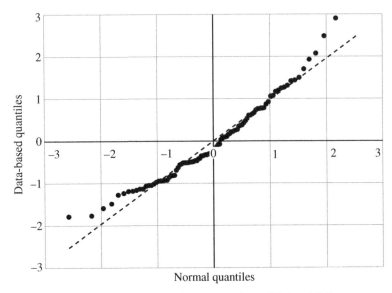

FIGURE 6.7   Q–Q plot of Santa Monica celebrity sightings.

As useful as it is, the Q–Q plot is only qualitative. Whether you determine your data to be normal or not is up to you and your subjective eye. Two different data analysts looking at the same plot might come to completely different conclusions. For this reason, there are a variety of hypothesis tests for normality that remove this subjectivity from the process.

Tests for normality compare two competing claims:

H_0 : Data conform to a normal distribution   vs.

H_A : Data do not conform to a normal distribution.

These two hypotheses are hardly mathematically precise. What does it mean for a dataset to conform to the normal distribution? Different statisticians have had different ideas how to answer this question over the years, and this had led to the development of a number of different tests for normality.

Moment-Based Tests for Normality   All probability distributions have unique characteristics, and the normal distribution is no different. This distribution is symmetric, the left and right halves of the pdf are mirror images of one another. It's also mound-shaped, downward sloping on either side of the midpoint with just the right amount of bulge to it. Any dataset that conforms to the normal distribution should have these characteristics, and looking at these characteristics is one way to test for normality.

The degree of symmetry and the amount of bulge in a probability distribution can be measured using moments. **Moments** are expected values describing the properties of a probability distribution. The mean and variance are examples of moments. The mean, $\mu = E[X]$, describes the arithmetic center of a distribution and the variance, $\sigma^2 = E[(X - \mu)^2]$, describes the variation around the mean. There are other moments that describe other properties of a distribution, and two of these are particularly important when it comes to tests for normality. **Skewness** measures symmetry. Mathematically, skewness is the third central moment: $\text{Sk} = E[(X - \mu)^3]$. If, like the normal distribution, a distribution is perfectly symmetric, then $\text{Sk} = 0$. **Kurtosis** measures the amount of bulge a distribution has. Mathematically, the kurtosis is the fourth central moment, or $K = E[(X - \mu)^4]$. The kurtosis of the normal distribution with variance σ^2 is $K = 3\sigma^4$.

The **D'Agostino Pearson test** uses skewness and kurtosis to test for normality. Specifically, if a dataset conforms to the normal distribution, then the sample skewness and kurtosis, calculated from the data values, should approximately equal the skewness and kurtosis of a normal distribution with the same mean and variance. This logic leads to the following hypotheses:

$$H_0 : \text{Sk} = 0 \text{ and } K = 3\sigma^4 \qquad \text{vs.}$$
$$H_A : \text{Sk} \neq 0 \text{ or } K \neq 3\sigma^4$$

The mathematical details of this test are beyond the scope of this book. However, like any hypothesis test, the D'Agostino Pearson test uses a test statistic, this one calculated from the sample skewness and kurtosis, to calculate a p-value, and compares that p-value to a critical threshold based on a significance level you specify.

The D'Agostino Pearson test for normality appears frequently in data analysis software packages. You provide a data vector and a significance level and the software does the rest. This test is one of the original tests for normality, but it isn't the most commonly used. Newer tests have been developed that are more powerful, meaning they're able to detect more subtle deviations from normality with the same significance level. I don't use the D'Agostino Pearson test in this book, but I've included this summary because when you're faced with data analysis software that offers a half dozen different tests for normality, it's nice to have some idea of the differences between them.

Goodness-of-Fit Tests for Normality   Recall the Q-Q plot. The Q-Q plot is a plot of the quantiles of a dataset against the quantiles of a corresponding normal distribution. If the data conform to a normal distribution, then the points in a Q–Q plot should fall on a nice straight line. Most modern tests for normality use some variation of quantiles to determine if the differences between a sample and the normal distribution are statistically significant. Because they compare empirical and theoretical probability distributions across a range of values, these tests fall under the general category of goodness-of-fit tests.

The mathematical derivation of the typical modern test for normality requires some statistical sophistication, and so details are omitted here. The motivated reader can find more information in books such as in Sheskin, *Handbook of Parametric and Nonparametric Statistical Procedures*. Fortunately, most data analysis packages offer at least one test for normality, so it's not absolutely necessary to know the details of the tests to use them. However, knowing the similarities and differences between the most popular tests can help you interpret your results, especially if you find different tests coming to different conclusions about your data.

Three of the most common tests for normality are the Kolmogorov–Smirnov test, the Anderson–Darling test, and the Shapiro–Wilk test. The first two of these tests can be used to compare a sample to any distribution, not just the normal distribution. This can be done in many data analysis packages by specifying which distribution you'd like to test against. In this chapter, I'll restrict my attention to testing for normality.

The Kolmogorov–Smirnov test is a **nonparametric** test. Nonparametric methods make no assumptions about the underlying probability distribution of a sample. This means they can be applied to all kinds of data without worrying about normality. It also means this test can be used to test not only for normality but also for any other probability distribution you'd like. To do this, the Kolmogorov–Smirnov test compares the empirical cdf, calculated from the sample quantiles, to the theoretical cdf of the desired probability distribution, and then finds the largest deviation between them. If this largest difference is bigger than what you'd expect purely by chance, the null hypothesis of normality is rejected.

The Anderson–Darling test and the Shapiro–Wilk test are based on **order statistics**, observations that have been sorted, or ordered, from smallest to largest. The **kth-order statistic** is the kth smallest observation. In other words, if you sort a sample from smallest to largest, the kth value in this sorted list is the kth-order statistic. Where the original observations are typically written as $X_1, X_2, \ldots, X_N$, the order statistics are written with a parentheses around the subscript, $X_{(1)}, X_{(2)}, \ldots, X_{(N)}$ to indicate they've been reordered.

Like quantiles, the order statistics for a sample can be compared to those for a normal distribution to test for normality. Both the Anderson–Darling and the Shapiro–Wilk test use a test statistic calculated from the sorted values along with known properties of order statistics for the normal distribution. But it's not the same test statistic. The Anderson–Darling test statistic can be

Test	Test statistic	Properties
D'Agostino–Pearson	Based on moments	• One of the first tests for normality. • Not as powerful as more modern tests for normality
Kolmogorov–Smirnov	The maximum difference between the sample and normal quantiles	• Commonly used • Nonparametric • Can be used to test against any probability distribution
Anderson–Darling	Based on order statistics	• One of the more powerful tests for normality • Can be used to test against the normal, uniform, exponential, lognormal, Weibull, and extreme value distributions
Shapiro–Wilk	Based on order statistics	• One of the most powerful tests for normality • Specific to the normal distribution

FIGURE 6.8   Tests for normality and their properties.

used to test whether a sample conforms to a variety of probability distributions, whereas the Shapiro–Wilk test statistic is specifically tailored to the normal distribution.

The Shapiro–Wilk test is one of the most powerful tests for normality, followed by the Anderson–Darling test and the Kolmogorov–Smirnov test. Figure 6.8 summarizes this and other key properties of the various tests for normality.

All tests for normality have their strengths and weaknesses. When determining if a sample conforms to the normal distribution, it's usually a good idea to generate a Q–Q plot and run at least one of the tests and compare the two. For example, when applied to the celebrity sightings data, the Anderson–Darling test for normality produces a p-value of 0.008. This is well below the typical significance level of $\alpha = 0.05$, and supports my suspicions from the Q–Q plot in Figure 6.7 that the data are right-skewed. Therefore, I'd conclude these data are most definitely not normal.

DATA TRANSFORMATIONS AND OTHER STRATEGIES FOR COPING WITH NON-NORMAL DATA

When you find yourself with a sample that isn't normal, there are three things you can do. First, you can ignore the non-normality and proceed with a statistical analysis based on the normal distribution. This strategy works when you have a large sample ($N \geq 25$) and when you're constructing confidence intervals or testing the mean or a proportion of your population.

The second strategy for dealing with non-normal data is to find a statistical technique based on a probability distribution that better suits your data. There are techniques based on virtually any probability distribution you can imagine, and a quick Internet search will usually uncover them. If you don't want to go to the trouble of identifying a probability distribution to fit your data, you can always use nonparametric techniques. Nonparametric techniques make no assumptions about the probability distribution of your data, and they're usually robust when it comes to oddities like outliers and funky looking histograms. A variety of common nonparametric techniques are introduced later in this book and I refer you to Chapter 8 for more information on this topic.

Finally, in some cases, a data transformation can be used to make your data more normal. A **data transformation** is just a function you apply to all your data values. Data transformations are incredibly useful. They're simple, and they allow you to use the multitude of techniques designed for the normal distribution. The most common type of non-normal data are right-skewed data. Certain types of observations—distances, areas, counts of people or items, dollar values, and measurements ranging over many orders of magnitude—tend to have this property. The **log transformation** works especially well with right-skewed data. To perform the log transformation, you take the logarithm of every data value and use those transformed observations in your analysis.

This approach works well for testing hypotheses with skewed data. Perform the test on the transformed observations and you're done. For estimates and confidence intervals, some data analysts construct the interval on the transformed data and then apply the reverse transformation to get those estimates back into the original units. However, doing this tends to produce biased estimates and shifted confidence intervals. I don't recommend this approach. But if you must use a log transformation to construct a confidence interval, it's a good idea to compare the result with the corresponding interval generated another way, for example, using the bootstrapping procedure introduced in the last chapter or, when you have a large sample size, the traditional confidence interval based on the normal approximation. The case study that follows provides an example of how transformed data can be used in a typical data analysis.

THERE'S NORMAL, AND THEN THERE'S HOLLYWOOD NORMAL

Hollywood celebrity marriages are anything but normal. Forget that the celebrities are paid to project an image that rarely reflects reality. Forget that many of them can't go to the store without a mob of paparazzi trailing behind them.

Forget that they're multimillionaires who'll never need to worry about paying their gas bill. Just the fact that there's a world of voyeurs out there who thrive on every juicy detail of their lives makes them so different from the average couple, it's almost impossible for the rest of us to imagine what it's like.

Conventional wisdom tells us Hollywood relationships never last beyond the honeymoon stage. But is this true? Are celebrity marriages really shorter than the average marriage? Are they really measured in days, not years?

The last place I wanted to go for answers was the supermarket tabloids. These newspapers specialize in scandal, so they'd hardly be inclined to report any long-lasting happy Hollywood marriages out there. Instead, I relied on two more impartial websites to collect my sample. The first website, called rottentomatoes.com, contains movie reviews and celebrity information. Every year, this website publishes the Rotten Tomato Awards, a list of the top ten movies of the year, as voted on by their large and growing list of Internet critics. To generate a sample of celebrities to scrutinize, I grabbed the names of the headliners from the Golden Tomato top ten lists for all years between 2009 and 2013. With this list of celebrities in hand, I visited another website, called imdb.com, which also specializes in all things Hollywood. This website contains detailed biographies of most major movie stars, including marriages and divorces. In the end, I had the birthdays, marriages, and divorces of over ninety celebrities, ranging from film moguls like Brad Pitt to fresher faces like Emma Watson.

The ages of the film stars in my sample ranged from under 20 to 80. I eliminated all celebrities under thirty, keeping only those individuals who've had a little time to build a marriage track record. Figure 6.9 summarizes my findings.

I guess our scandal-obsessed culture has biased my perceptions because I must admit I was surprised when I saw the results. A large number of the celebrities, over 60%, have never been divorced. Many are happily married. Others have simply never married at all. There are a number of celebrities

Percent of celebrities who've never been divorced:	62%
Longest marriage:	Thirty-three years (Leonard Nemoy)
Shortest marriage:	Seven months (Woody Harrelson)
% of over-50 celebrities who've been married longer than 15 years:	24%
% of over-40 celebrities who've been married longer than 10 years:	30%

FIGURE 6.9 Celebrity divorce facts.

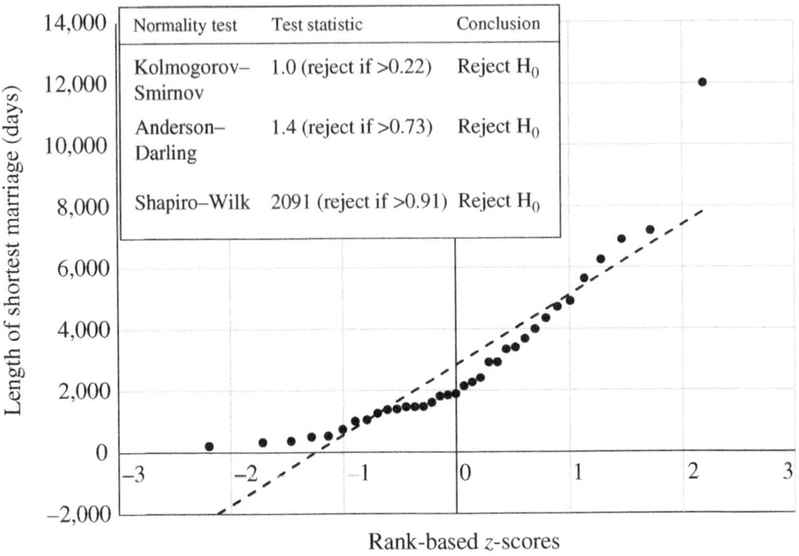
Normality test
Test statistic
Conclusion
Kolmogorov–Smirnov
1.0 (reject if >0.22)
Reject H_0
Anderson–Darling
1.4 (reject if >0.73)
Reject H_0
Shapiro–Wilk
2091 (reject if >0.91)
Reject H_0
Length of shortest marriage (days)
14,000
12,000
10,000
8,000
6,000
4,000
2,000
0
–2,000
–3
–2
–1
0
1
2
3
Rank-based z-scores

FIGURE 6.10 Q–Q plot of celebrity marriages.

who've had marriages lasting longer than ten years, including, among others, Julia Louise Dreyfus of *Seinfeld* fame, Leonard Nemoy from *Star Trek*, and Robert Downey, Jr., also known as *Iron Man*.

According to the United States Census Bureau, a typical American marriage lasts between fourteen and twenty years (twenty years for first marriages and fourteen years for second marriages). A confidence interval for the mean length of the celebrity marriages in my sample should tell me whether Hollywood marriages are normal or not. However, my observations are counts—number of days married—and they also vary by several orders of magnitude, ranging from 205 days to 12,002 days. Both of these things point to right-skewed data. So, before calculating any confidence intervals, I decided to check for normality. Figure 6.10 shows a Q–Q probability plot of the celebrity marriage data, with the results of three tests for normality overlayed on the graph. This Q–Q plot shows a characteristic tilted "U" pattern. These data are right-skewed. What is more, all three tests are in agreement. Celebrity marriages are not normally distributed.

A log transformation often helps data like these, so I took the logarithm of every observation and generated a new Q–Q plot, shown in Figure 6.11. The "U"-shape is gone, replaced by a different pattern. Apparently this pattern is not strong enough to be significant, however, because two of the normality tests, Anderson-Darling and Shapiro Wilk, call for accepting the null hypothesis. Since these two powerful and popular tests are in agreement, I took

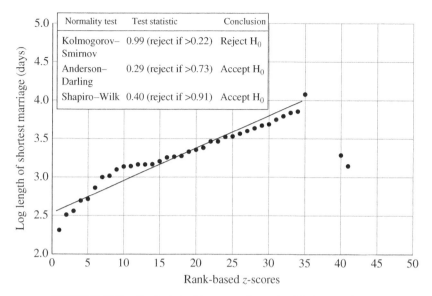

Normality test	Test statistic	Conclusion
Kolmogorov–Smirnov	0.99 (reject if >0.22)	Reject H_0
Anderson–Darling	0.29 (reject if >0.73)	Accept H_0
Shapiro–Wilk	0.40 (reject if >0.91)	Accept H_0

FIGURE 6.11   Q–Q plot of *log-transformed* celebrity marriages.

their recommendation and declared these log-transformed data to be sufficiently normal.

The estimated mean length of a marriage can be calculated from the log-transformed data by calculating the average and reversing the transformation. The average of the log-transformed data is 3.29. Raising this value to a power of 10 gives $10^{3.29} = 1944$ days or 5.32 years.

To construct a confidence interval for mean length of marriage from these log-transformed data, I calculated the upper and lower confidence bounds for the transformed data and then reversed the transformation. Specifically, the confidence interval for the log-transformed data is (3.15, 3.43). Taking these logged confidence bounds and reversing the transformation gives a confidence interval of ($10^{3.15}$, $10^{3.43}$), or (1404, 2693) days, or (3.8, 7.4) years.

How does this confidence interval compare to the traditional confidence interval calculated from the original data? Figure 6.12 compares the two. The original average is more than double the value derived from the log-transformed data. The confidence intervals overlap, but only just barely.

Which interval should I believe? Because I have more than twenty-five observations in my sample, the traditional confidence interval is probably pretty good. And because estimates based on log transformations tend to be shifted away from the true value, the disparity makes me suspicious of the confidence interval calculated from the transformed data. A bootstrap

	Original data (normal approximately)	Log-transformed data	Resample bootstrap
Average:	7.8	5.3	7.8
Confidence interval:	(5.4,10.2)	(3.8,7.4)	(5.6,10)

FIGURE 6.12   Comparison of confidence intervals constructed from original and log-transformed celebrity marriage data.

confidence interval (see Chapter 5) might add some insight and help me decide. As shown in Figure 6.12, a standard resample bootstrap confidence interval using 500 resamples and the original data produced a confidence interval of (5.6, 10.0) years. This agrees closely with the original confidence interval and not the interval based on the log-transformed data. Because of this, I'll go with the original estimate of 7.8 ± 2.4 years. And because this interval is so far below the U.S. Census Bureau's estimates of fourteen and twenty years, I can safely accept the stereotypes pushed by the supermarket tabloids. In other words, Hollywood marriages are definitely not normal.

BIBLIOGRAPHY

Elliott D. How Long do Marriages Last? United States Census Bureau, August 25, 2011.

Hollander M, Wolfe DA. *Nonparametric Statistical Methods*. New York: John Wiley & Sons, Inc; 1999.

Huff Post Shortest Celebrity Marriages. Available at http://www.huffingtonpost.com/2011/10/31/shortest-celebrity-marriages_n_1068180.html. Accessed November 2, 2011.

Jarman K. *The Art of Data Analysis: How to Answer Almost Any Question Using Basic Statistics*. Hoboken: John Wiley & sons, Inc; 2013.

NIST. Anderson-Darling Test. Available at http://itl.nist.gov/div898/handbook/eda/section3/eda35e.htm. Accessed May 24, 2014s.

NIST. Anderson-Darling and Shapiro-Wilk Tests. Available at http://itl.nist.gov/div898/handbook/prc/section2/prc213.htm. Accessed May 24, 2014b.

NIST. Kolmogorov-Smirnov Goodness-of-Fit Test. Available at http://itl.nist.gov/div898/handbook/eda/section3/eda35g.htm. Accessed May 24, 2014c.

Olsson U. Confidence intervals for the mean of a log-normal distribution. J Stat Educ 2005;13(1). Available at http://www.amstat.org/publications/JSE/v13n1/olsson.html. Accessed November 19, 2014.

Sheskin DJ. *Handbook of Parametric and Nonparametric Statistical Procedures*. 3rd ed. New York: Chapman & Hall; 2004.

Triola MF. *Elementary Statistics*. 11th ed. Toronto: Pearson; 2011.

7

BELIEVE IT OR DON'T: USING OUTLIER DETECTION TO FIND THE WEIRDEST OF THE WEIRD

Did you hear about the man who stole a GPS, but ended up having to call 911 when he got lost? How about the murder of nine college students that was recently blamed on the Yeti (the Russian Bigfoot)? Or the California school kids who made a fifty-foot long peanut butter and jelly sandwich in less than three minutes?

On any given day, you can search the Internet and find stories like these—true reports that are stranger than fiction. Many news outlets carry weird news stories, and a surprising number of these stories seem to come from the state of Florida. In fact, a recent Google search on "weird news Florida" revealed over 47,000,000 hits. That's a lot of strange. And from sewer-surfing alligators to whale-wrangling nudists, the sunshine state has it all.

Outliers are the Florida of the data analysis world. These strange observations sit at the extremes, far away from the rest of your data. And like the story about a twelve-foot python caught wrapping itself around an unsuspecting woman's toilet, outliers can leave you slightly disturbed, wondering what might've happened had they not been found. In this chapter, News of the Weird stories will be studied, and outlier detection will be used to find the weirdest of the weird.

Beyond Basic Statistics: Tips, Tricks, and Techniques Every Data Analyst Should Know,
First Edition. Kristin H. Jarman.
© 2015 John Wiley & Sons, Inc. Published 2015 by John Wiley & Sons, Inc.

THE WORLD OF THE WEIRD

If you've read the last few chapters, you've already run across outliers. You know they're extreme values that sit far away from the center of a dataset. You've seen how they can ruin the results of a statistical analysis. And you've learned there are techniques for minimizing the impact these weird values have. In this chapter, you'll learn some techniques for finding and eliminating extreme values before they have a chance to influence your analysis. If you're comfortable with the concept of outliers and you've seen how much damage they can do, then go ahead and skip to the next section. If you'd like to see one more illustration, then this section is for you.

It only takes one or two extreme values to shift a sample mean, inflate a standard deviation, and bias a slope estimate. Consider the two datasets in Figure 7.1.

These data are **simulated**, made up, and just about as perfect as data get. Almost. There are twenty observations in Group 1, and all of them are normally distributed observations with mean 5.5 and variance 1. There are also twenty observations in Group 2. Eighteen of them are normally distributed with mean 6 and variance 1. The remaining two, the extreme values at 1.8 and 2.1, are outliers. Figure 7.2 shows what these two outliers can do to some basic statistics.

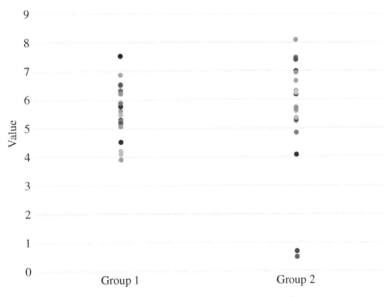

FIGURE 7.1   Normal data with two outliers.

Summary statistics:

	Group #1	Group #2 (with outliers)	Group #2 (without outliers)
Sample mean	5.52	5.54	6.10
Standard deviation	0.95	1.95	1.01

T-test results:

p-value with outliers	p-value with outliers
0.49	0.04

FIGURE 7.2   The impact of outliers on common statistical techniques.

The sample mean and standard deviation of Group 1 are close to the respective values of 5.5 and 1.0 that we know to be true. The sample mean and standard deviation for Group 2 are not close to the true values. Where the sample mean should be about 6, it's 5.68. Where the standard deviation should be close to 1, it's 1.6. Removing the two obvious outliers dramatically improves the sample statistics, increasing the sample mean to 6.1 and decreasing the sample standard deviation to 1.01.

The impact of these outliers on a t-test is even more dramatic. Because the standard deviation is inflated when the outliers are included, the p-value is a statistically insignificant value of 0.49, and we'd be forced to accept the null hypothesis that the means are the same, even though they're different. When the two outliers are excluded, however, the p-value drops below the 0.05 significance level, leading to the correct conclusion that the means of the two groups are different.

OUTLIER AND ANOMALY DETECTION: KNOWING AN ODDITY WHEN YOU SEE ONE

In this age of easy data collection, outlier detection is an important part of any thorough data analysis. Outliers can be mistakes, such as transposition or copy-and-paste errors. They can be faulty measurements. Or they can be perfectly legitimate, but strange data values. It's important to hunt down and correct any errors so you know your analysis is accurate. It's also important to take note of legitimate weird values because where there's one, more could easily follow.

Correcting mistakes and understanding your data are only two reasons why outlier detection is important. In some cases, you might be more interested in the unusual values than the typical values. When this is the focus of an analysis, finding and flagging unusual values, it's called **anomaly detection**.

Anomaly detection is used in many modern applications. For example, credit card companies often track spending patterns of their customers to monitor for fraud, and they use sophisticated anomaly detection methods to do it. Each customer has his or her unique spending pattern. When spending becomes extreme or unusual, the anomaly detection programs raise a red flag, and the company suspends the card until it can get confirmation that the purchases are legitimate. Anomaly detection pops up in other fields, too. Computer scientists use it to monitor Internet traffic for scams, spam, and viruses. Manufacturing engineers use it to look for defects in products coming off a production line. Financial analysts use it to look for blips in the stock market that might foretell good (or bad) things to come.

There's a difference between outlier detection methods and anomaly detection methods, sort of. Both types of methods look for unusual observations that don't fit with the rest, but most outlier detection methods assume outliers are rare, only a few per dataset, and that they are most likely erroneous. Anomaly detection methods, on the other hand, tend to allow for a larger number of unusual observations, and they also allow that these anomalies are perfectly valid, albeit unusual values. Regardless of whether you come from the outlier detection camp or the anomaly detection camp, the most common, most basic techniques for finding them are the same.

The simplest techniques for finding outliers are graphical. A good plot can reveal all kinds of information about your data, including extreme and unusual values. The simple scatter plot like the one in Figure 7.1 can instantly reveal odd, extreme values. A bar chart or histogram (Chapter 5) and a Q–Q plot (Chapter 6) can also reveal outliers. But these techniques don't always work in every situation. Sometimes, you need more than a good graph.

z-scores, Not Just for t-Tests Anymore

Suppose you're monitoring parts coming off an assembly line, or watching Internet traffic pass through your website. You want to look for outliers, but there's no time to stop and generate box plots every time you get a new observation. In cases like these, you need an outlier detection technique that can run on its own. The simplest such technique relies on one of the most basic of all statistics: the z-score.

You may recall the z-statistic. If you have a random variable X drawn from a normal distribution with mean μ and variance σ^2, the value

$$z = \frac{X - \mu}{\sigma}$$

is a **standard normal** random variable, meaning a random variable with a normal distribution having mean $\mu=0$ and variance $\sigma^2=1$. A **z-score** is a sample-based version of the z-statistic. For any observation x_i from a dataset having sample mean $\overline{x}$ and sample standard deviation s, the corresponding z-score is

$$z_i = \frac{x_i - \overline{x}}{s}.$$

The z-scores have one incredibly useful property. If your sample is drawn from a normal distribution, then the z-scores have a standard normal distribution. In other words, it doesn't matter what the original sample mean and variance are. As long as you have a reasonably bell-shaped dataset, you can **standardize**, or transform, the observations to a distribution having many nice properties. When it comes to outlier detection, one of the nicest of these properties is illustrated in Figure 7.3, specifically, about 99% of the z-scores in a well-behaved dataset should fall within the range ±2.5. This property makes it easy to hunt for outliers. You simply look for z-scores that are outside the interval −2.5 to 2.5. For example, Figure 7.4 plots the z-scores of the simulated data from Figure 7.1. Lines at −2.5 and 2.5 have been added for reference. The Group 1 z-scores all lie within the −2.5 to +2.5 range. No outliers there. Two of the Group 2 z-scores fall outside this range. These are the weird ones.

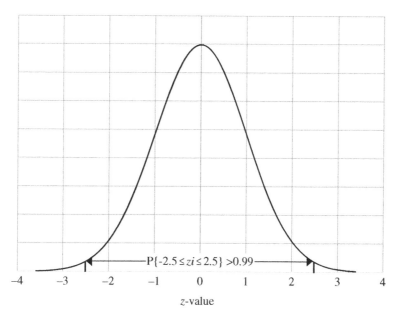

FIGURE 7.3 Interpreting z-scores.

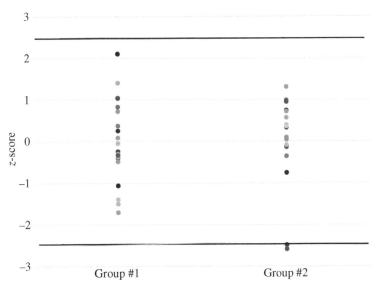

FIGURE 7.4   *z*-Scores of grouped data.

z-scores are widely used in outlier detection, particularly in quality control of laboratory or manufacturing processes. When combined with the appropriate graph, this approach can be an effective way to flag strange or defective items. However, there are two problems with this method. First, *z*-scores are calculated using the sample mean and standard deviation, two statistics that can be dramatically impacted by extreme values. If you have a small sample or more than just a few outliers, your sample mean might be skewed and your standard deviation might be badly inflated. A skewed sample mean can cause skewed *z*-scores, which can push perfectly good observations outside the acceptable −2.5 to 2.5 range. A badly inflated standard deviation can cause badly shrunken *z*-scores, which can pull outliers into the acceptable −2.5 to 2.5 range. In other words, the *z*-score approach to identifying outliers works well, as long as your data are roughly bell-shaped and only a small percentage of the observations are outliers.

The Interquartile Range Test for Robust Outlier Detection

Are there outlier detection techniques that aren't impacted by outliers? Yes, and they're based on robust statistics (see Chapter 5). Robust statistics are specifically designed to be insensitive to oddities in the data. The median is a robust statistic. Up to half of the values in a sample can be extreme, and the median won't be impacted in any way.

Recall from Chapter 5, the interquartile range (IQR), or middle fifty, is the distance between the 75th percentile and the 25th percentile in a dataset. In other words, it's the range inside which the middle 50% of observations lie. The IQR measures the variation around the center location, and for bell-shaped data, this statistic is about two-thirds of a standard deviation. The breakdown point of the IQR is 25%, making it significantly more resistant to outliers than the more traditional standard deviation.

The simplest robust outlier detection method relies on the first and third quartiles and the IQR. If Q_1 is the first quartile, the 25th percentile, and if Q_3 is the third quartile, the 75th percentile, you simply look for values that are outside the following range:

$$(Q_1 - 1.5\text{IQR}, Q_3 + 1.5\text{IQR}).$$

Figure 7.5 shows the grouped data from Figure 7.1 with lines indicating this range. Where the z-scores method shows the two known outliers as being close to the threshold for typical observations, this test has both of them far outside the acceptable range for normal observations.

Rules-of-thumb-based outlier detection schemes like these are effective and simple to implement. However, there are more sophisticated methods, most of which are based on formal hypothesis tests. These aren't as simple

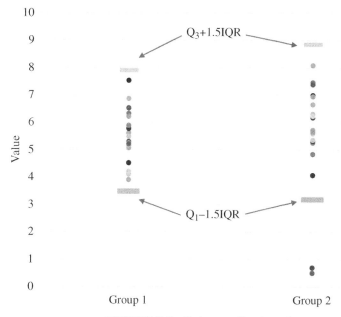

FIGURE 7.5   Robust outlier detection.

as rule-of-thumb methods because when testing extreme values, you can't rely on the Central Limit Theorem to give you a nice, normal distribution to work with. As a result, many formal outlier detection methods require specialized probability distributions to calculate critical values, and you won't find most of them in a typical statistics book. They do have their benefits, though, so I've included one such technique here.

Grubbs' Test: An Exact Test for Finding Outliers

The z-score and IQR outlier detection methods are based on well-known rules of thumb, and so these are approximate tests. **Grubbs' test** is a formal hypothesis test, complete with a test statistic, a critical threshold, and error probabilities. The hypotheses for Grubbs' test are as follows:

H_0 : There are no outliers in the dataset vs.

H_A : At least one outlier is present.

Grubbs' test assumes the data follow a normal distribution, at least under typical, nonoutlier circumstances, and so not surprisingly, the test statistic is based on z-scores. It's very different from the z-score method for detecting outliers, though. Where the z-score test simply flags individual data values whose z-scores fall outside the range of -2.5 to 2.5, Grubbs' test uses something called an **extreme value distribution**. An extreme value distribution is a probability distribution specifically designed to describe the largest (or smallest) value in a set of data. Extreme value distributions are important because even when the original observations are nicely normal, the probability distribution of the extreme values is most decidedly non-normal. By relying on a precise probability distribution rather than z-scores and a rule of thumb, Grubbs' test provides the confidence that only a formal hypothesis test can give.

Grubbs' test statistic is easily calculated from the z-scores. Specifically, if $z_1, z_2,$ and so on are the z-scores, Grubbs' test statistic is

$$G = \max\left(|z_1|, |z_2|, \ldots |z_N|\right).$$

In words, G is the absolute value of the most extreme z-score. For example, of the Group 2 z-scores plotted in Figure 7.4, the most extreme value is -2.60. Grubbs' test statistic would then be $G = |-2.6| = 2.6$.

Unfortunately, working with extreme value distributions can be messy. The decision criterion for Grubbs' test is a complicated formula involving critical values for the t-distribution. Specifically, you reject the null hypothesis if

$$G > \frac{N-1}{\sqrt{N}} \sqrt{\frac{t^2_{(\alpha/2N),N-2}}{N-2+t^2_{(\alpha/2N),N-2}}}$$

where $t_{(\alpha/2N),N-2}$ is the $\alpha/2N$ critical value for the t-distribution with $N-2$ degrees of freedom. Critical values for Grubbs' test are tabulated in Appendix D. To use this table, simply choose a significance level, say 0.05, and look up the critical value corresponding to the appropriate sample size. For the Group 2 data, $N=20$. Therefore, the critical value for Grubbs' test with $\alpha=0.05$ and $N=20$ is $G_{crit}=2.83$. Since $G=2.6$ is less than this critical value, you'd accept the null hypothesis that there are no outliers. In other words, Grubbs' test does not find the two known oddities in this dataset.

Previously, we saw that because it's robust, the IQR test tends to be a little more sensitive than the z-score method, meaning it tends to find outliers a little easier. Grubbs' test suffers from the same problem the z-score method has, namely, when the standard deviation is inflated due to extreme values, the z-scores will tend to be smaller than they should be. This can reduce the power, the probability of finding outliers, of Grubbs' test. On the other hand, because you specify a significance level, Grubbs' test guarantees a certain Type I error probability. So you can be sure the probability of flagging false outliers is low. Neither of the other two tests provides this level of confidence in the results.

Grubbs' test doesn't flag individual outliers. And because the test statistic is based only on the most extreme z-score, this test is only good for detecting one outlier. In other words, if the null hypothesis is rejected and you conclude you have an outlier, you only know the most extreme value is an outlier. You don't know anything about the second or third most extreme values.

It's somewhat common practice to modify Grubbs' test to identify multiple outliers as follows. Perform the test on the entire dataset. If the null hypothesis is rejected, then remove the most extreme value and repeat the test with the remaining measurements. Do this until the null hypothesis is accepted. This approach may seem reasonable, but if you choose to do this, be aware that applying a test, any test, to a dataset over and over again increases the error probabilities you specify for the test. A generalized version of Grubbs' test is available, where you can test for up to r outliers at once. This test, called the **generalized extreme residual test**, adjusts the significance level to account for the increase in error caused by repeatedly applying Grubbs' test. A description of this test can be found at http://itl.nist.gov/div898/handbook/eda/section3/eda35h3.htm.

SO, YOU'VE FOUND AN ODDITY. WHAT NOW?

Suppose you look for outliers by looking for z-scores that fall outside the range of -2.5 to 2.5. According to the standard normal distribution, the probability any given z-score falls outside this range is 0.01 or less. In other words, the probability a perfectly normal observation will be flagged as an outlier is about 0.01, or 1%. This means you can expect about one out of a hundred perfectly good observations to be labelled outliers. This is manageable when you only have a hundred or so observations, but many modern data analyses involve thousands or millions of observations. With a thousand data values, you could expect ten falsely flagged measurements. With ten thousand data values, you could expect 100 false outliers. In other words, for large volumes of data, even though the error rate is small, the number of potential outliers you end up tracking down can grow to be quite large.

I'm not just picking on z-scores. This phenomenon isn't restricted to rule-of-thumb-based outlier detection methods. This is a problem for any tests that is repeated over and over in a single data analysis. Each time you repeat the test, there's a small probability you'll declare false significance. If you only do that a few times, it's not a big concern. However, if you repeat it many hundreds or thousands of times, the number of falsely flagged significances grows rapidly. This problem is common in big data applications such as computer security, where millions of network transmissions need to monitored, and in airport security technologies, where huge volumes of passengers are being checked for forbidden items. And this phenomenon is so notorious in bioinformatics, where data analysts routinely screen tens of thousands of genes in a single analysis, the researchers in this field have given it a name: **false discovery**.

Several techniques have been developed for minimizing the false discovery rate. In particular, if you're running an outlier detection procedure, it would be nice to know that the entire process, not each individual test, has a probability α of finding one or more false outliers. A **Bonferroni adjustment** is one way to guarantee this. A Bonferroni adjustment is made by dividing the significance level α into the total number of tests you plan on running. For outlier detection, this means setting the significance level of each comparison to α/N. Unfortunately, a decrease in α gives rise to a corresponding decrease in the sensitivity, or power of the test. So, when you have more than just a few tests to run, the Bonferroni adjustment reduces the number of false significances found, but it also reduces the probability that true outliers will be detected.

Other, more sophisticated methods for reducing the false discovery rate are available, but in the end, not much can be done about this problem. If you have a big dataset and many tests to run, it's a limitation you learn to live with. Because of issues like this, you can't simply throw out an outlier just because you find one. Outlier (or anomaly) detection is only the first part of the analysis. If you find an erroneous value, you need to investigate it. Is there a simple explanation such as cut-and-paste error? If it's a scientific instrument, did you forget to calibrate it or did experimental conditions make it a suspect value? If you're doing anomaly detection, monitoring a stream of data, is correlation to blame for a few extreme values? Are your measurements drifting due to uncontrollable changes in the environment?

With all outliers, it's important to examine each one and try to understand the cause of it. If you find an error or mistake, and if it can't be fixed, then it's usually OK to remove the value from the dataset and forget about it. If you can find no explanation as to why the value is so unusual, then it might just be a perfectly legitimate value. And where there's one, there's likely to be more. So when you have a legitimate extreme value such as this, it might be best to leave it in the analysis.

THE WEIRDEST OF THE WEIRD

Wyoming is full of gun-toting cowboys. California is the land of face lifts and road rage. New Hampshire is filled with peace-loving ice cream fanatics. And Florida? Well, Florida is just weird.

Or is it?

There's no better place to go for an answer to this question than Chuck Shepard's popular website, News of the Weird. Mr. Shepard, the Czar of strange news, has been researching and writing about the unusual since the 1970s. And his website is a gold mine of reports involving strange human behavior. My plan? To gather a random sample of news reports from the News of the Weird website, identify the location in which each of these stories occurred, tally them up by state, and use outlier detection to find the weirdest of the weird.

There are thousands of stories on the News of the Weird website. With fifty states (no US territories or District of Columbia used here), I wanted to make sure I got at least a handful of reports from most of them. So, I randomly chose 200 issues between 2011 and 2014, and began tallying up the weird news by state.

There really is such a thing as too much information. Reading this many news stories about deviant human behavior was an education, and not one I'm sure I wanted. But I stuck with it. Leaving out those reports that didn't specify a location and those that occurred outside the fifty US states, I ended up with 735 stories, or observations in my sample.

Figure 7.6(a) plots the total number of weird news stories by state for all states listed in alphabetical order. The hands-down winner in the weird news department

(a)

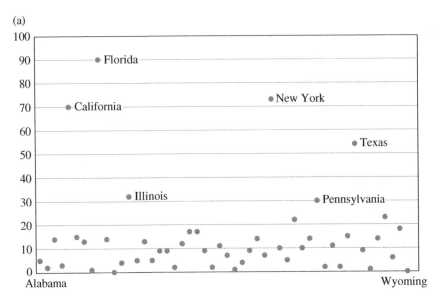

(b)

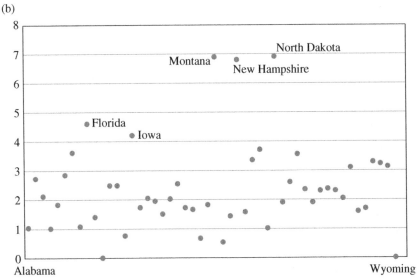

FIGURE 7.6   (a) Total Number of Weird News Stories by State (2011–2014), (b) Number of Weird News Stories per 1,000,000 Residents by State (2011–2014).

was Florida with ninety stories, followed by New York with seventy-three stories, California with seventy stories, and Texas with fifty-four stories. Hawaii and Wyoming had no stories. Every other state was somewhere in between.

Florida, New York, California, and Texas had by far the most stories, but they're also the four most populous states as well. More people, more opportunities for weirdness. To minimize the impact of population on my analysis,

I normalized the weird news data by the number of residents in each state. Figure 7.6(b) plots the number of weird new stories per 1,000,0000 residents, as reported by the US Census Bureau. After taking population size into account, New York, California, and Texas disappear into the cloud of data. Florida still remains high, but it's significantly lower than three much less populous states: North Dakota, New Hampshire, and Montana.

Figure 7.7 displays a boxplot of the number of weird news stories per 1,000,000 population. According to the boxplot, the median rate of weird news is about two per 1,000,000 people. The IQR spreads from about 1.5 to 3 per million. The minimum is zero, and the maximum is seven. The long whisker stretching from the 75th percentile to the maximum suggests either a heavily skewed distribution or outliers.

Figure 7.8 displays the results of the z-scores, IQR, and Grubbs' test outlier detection methods applied to the weird news rate per 1,000,000 residents. As before, Grubbs' test isn't sensitive enough to identify outliers in this dataset. However, both the z-scores and the IQR method flagged North Dakota, New

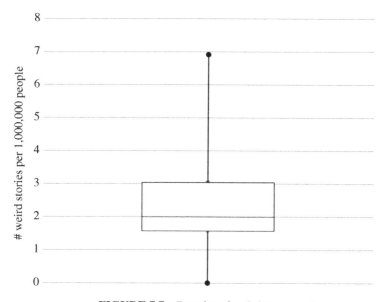

FIGURE 7.7   Boxplot of weird news stories.

Technique	Results	Weird states
z-Scores	Three outliers found	ND, MT, NH
IQR	Three outliers found	ND, MT, NH
Grubb's test	Accept H_0	None

FIGURE 7.8   Is Florida the weirdest of the weird?

Hampshire, and Montana as weird states. Not because they had a lot of reports, but because these states have such low populations, just a few weird news stories can launch them into the extremes. Of the four most heavily populated states, Florida was closest to being flagged an outlier by both rule-of-thumb methods. This state had 4.6 weird news stories per 1,000,000 people. The upper threshold for the IQR method is 5.2, well above this value. Additionally, a 4.6 per one million weird news rate translates to a z-score of 1.6, well below the 2.5 cutoff value for outliers using this criterion.

Florida does have a lot of weird news, more than any other state. However, according to the outlier detection methods presented in this chapter, the difference isn't even close to being statistically significant. The residents of the sunshine state may well be the weirdest of the weird, but they need to work a lot harder if they want to prove it.

BIBLIOGRAPHY

Brennan J. Detecting Outliers Using Z-Scores in Excel. Available at http://www.itl.nist.gov/div898/handbook/eda/section3/eda35h.htm. Accessed March 7, 2014.

Dawson R. How significant is a boxplot outlier? J Stat Educ 2011;19 (2):1–13.

National Institute of Standards and Technology. Engineering Statistics Handbook. Available at http://www.itl.nist.gov/div898/handbook/eda/section3/eda35h.htm. Accessed March 10, 2014.

Rousseeuw P, Leroy A. *Robust Regression and Outlier Detection*. 3rd ed. Hoboken: John Wiley & Sons, Inc; 1996.

Shepard's C. News of the Weird Archive. Available at http://newsoftheweird.com/archive/index.html. Accessed May 31, 2014.

Sheskin DJ. *Handbook of Parametric and Nonparametric Statistical Procedures*. 3rd ed. New York: Chapman & Hall; 2004.

The Huffington Post. Huff Post Weird News: Strange but True Stories from the Headlines. Available at http://www.huffingtonpost.com/weird-news/. Accessed June 7, 2014.

Triola MF. *Elementary Statistics*. 11th ed. Toronto: Pearson; 2011.

United States Census Bureau. National, State, and Puerto Rico Commonwealth Totals Datasets: Population, population change, and estimated components of population change: April 1, 2010 to July 1, 2013. Available at http://www.census.gov/popest/data/national/totals/2013/NST-EST2013-alldata.html. Accessed May 31, 2014.

Zap C. Why so Much Weird News Comes from Florida. Yahoo! News. Available at http://news.yahoo.com/-why-so-much-weird-news-comes-from-florida--143744802.html. Accessed July 31, 2013.

8

THE BATTLE OF THE MOVIE MONSTERS, ROUND TWO: RAMPING UP HYPOTHESIS TESTS WITH NONPARAMETRIC STATISTICS

In the 1962 classic science fiction movie, *King Kong vs. Godzilla*, two of the world's most famous movie monsters fight an epic battle. The prize? Bragging rights and the opportunity to terrorize the poor citizens of the small island of Japan. The battle lasts only a few minutes, but the giant ape and the fire-breathing lizard do plenty of damage in that time, toppling buildings and causing earthquakes and mudslides. In the end, locked in one final death grip, the two creatures roll off a cliff and into the sea. Kong is spotted swimming toward his home on Skull Island. Godzilla isn't seen again.

Was Kong running away from the nuclear lizard? Or did Godzilla sink to the bottom of the ocean in defeat? Who won the battle, anyway?

The producers leave room for debate over these questions, and it's a debate that rages on. Over fifty years after the movie's release, arguments over the outcome of the epic battle can be found in blogs and forums across the Internet. Godzilla fans insist the ape's strength was no match for the lizard's atomic breath and impenetrable scales. Kong fans insist the ape outwitted the lizard, leaving him to die at the bottom of the ocean. All of the fans are certain of their position. And yet, nothing has ever been proven.

Beyond Basic Statistics: Tips, Tricks, and Techniques Every Data Analyst Should Know,
First Edition. Kristin H. Jarman.
© 2015 John Wiley & Sons, Inc. Published 2015 by John Wiley & Sons, Inc.

In *The Art of Data Analysis: How to Answer Almost Any Question Using Basic Statistics*, Godzilla and King Kong engage in a very different kind of battle, this one to determine which classic monster is more popular. Using top ten movie monster lists collected from the Internet, the two creatures' rankings are compared using a *t*-test and a test for a proportion. Unfortunately, like the original classic movie, these standard hypothesis tests fail to come up with a clear winner.

And so the debate continues.

In this chapter, the movie monsters enter round two of the popularity battle. The weapon of choice in this second round: nonparametric hypothesis tests. Where standard hypothesis tests assume the data are approximately normally distributed, nonparametric methods make no assumptions about the underlying distribution of the data. Can Wilcoxon signed rank and other nonparametric methods declare a winner in the battle of the movie monsters?

THE PARAMETRIC HYPOTHESIS TEST: A CONVENTIONAL WEAPON FOR CONVENTIONAL BATTLES

The *t*-test, the *F*-test, and analysis of variance (ANOVA) assume your data follow the normal distribution. Common tests for a proportion assume either (i) your data follow a binomial distribution, or (ii) you have enough samples to use the normal approximation to the binomial distribution. Hypothesis tests like these, ones that assume the data have a specific underlying probability distribution, are called **parametric tests**.

Most conventional techniques, those you learn in a first statistics course, are parametric tests, and there's a good reason for this. Parametric tests usually work well. Many types of data naturally conform to the normal distribution, and there are a handful of normal approximations that can be applied to those that don't. As a result, tests based on the normal distribution usually produce reliable and accurate results. Usually, but not always. Suppose, for example, you had a sample whose frequency distribution looked like the one in Figure 8.1. This is an extreme example of a bimodal distribution. A **bimodal distribution** is a frequency distribution with two centers, where observations are clustered around each of these centers. In this case, the two centers are roughly located at −1 and 5.

A bimodal frequency distribution like this often shows up when you have observations representing two different phenomena. For example, suppose you were collecting data from customers shopping at a particular online book store. The search history of these customers would indicate

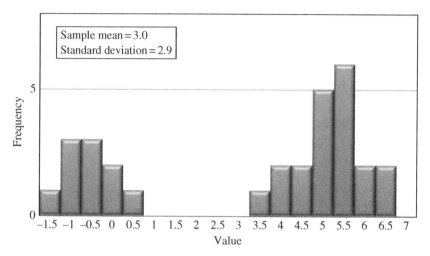

FIGURE 8.1 A bimodal frequency distribution.

what type of books they like to read, whether fiction, nutrition, self-help, or something else. If you were to count the number of self-help books browsed by each customer, the resulting data would probably have a bimodal distribution, with one sharp spike at zero representing those customers who aren't interested in self-help books, and another cluster of positive values representing those customers who are.

For bimodal data like those in Figure 8.1, the typical descriptive statistics can be misleading. If you were to look only at the sample mean, for example, without plotting the values, you might assume the data are roughly bell-shaped and centered around three. Not so. What's more, blindly applying the t-test to these data, where

$$H_0 : \mu = 0 \quad \text{vs.}$$
$$H_A : \mu \neq 0,$$

produces a p-value of 0.20, well above a typical 0.05 critical threshold for rejecting the null hypothesis. In other words, a t-test applied to these data would lead to accepting the null hypothesis that the population mean is zero. However, neither of the two clusters contributing to this dataset are centered on zero.

Strongly non-normal data such as these can be difficult to work with. The normal assumption doesn't apply, and there's no obvious probability distribution you can assume. Fortunately, there's a class of hypothesis tests that can be used in any situation, without regard to underlying probability distributions. These are called nonparametric tests.

NONPARAMETRIC TESTS: WHEN YOU NEED TO EXERCISE THE NUCLEAR OPTION

In *King Kong vs. Godzilla*, the two monsters didn't end up fighting by chance. Godzilla, the long-time arch enemy of the good people of Japan, had just returned for another romp through the already ravaged country. Guns and tanks couldn't kill the giant lizard. Fire didn't work, either. In a last ditch effort, the Japanese government decided to fly King Kong in from his tiny Skull Island home in the hopes the two monsters would destroy one another.

When your data don't follow any distribution you know, you still have one last option. There's a class of statistical techniques called nonparametric methods. Also called distribution-free methods, **nonparametric methods** don't assume your data follow a normal distribution. In fact, they don't make any assumptions at all about the shape of your data. Like parametric tests, nonparametric tests are hypothesis tests. In other words, two hypotheses are constructed, a test statistic is calculated, a decision threshold is determined. If the test statistic is below the threshold, you accept the null hypothesis. If it's above the threshold, you reject it in favor of the alternative. But where parametric tests use specific probability distributions to do this, nonparametric tests rely on properties that apply to all data.

Nonparametric methods are the nuclear option when it comes to hypothesis tests. You can use them on virtually any type of data, without regard to outliers, the shape of the frequency distribution, or even whether the data are discrete or continuous. As long as you can order your observations from highest to lowest, you can apply nonparametric statistics to it. Plus, there's a nonparametric alternative for just about any parametric test you can name— the *t*-test, the *F*-test, ANOVA, and so on.

Most nonparametric statistical tests convert the original numeric observations to signs or rankings. **Signs** (+ or −) are labels given to each observation based on whether or not they're greater (+) or less (−) than some reference value. For example, if you were a teacher and you wanted to test whether or not the median grade on your final exam was a *C*, you could go through the letter grades of each student and assign them a sign, (+) for all the *A*s and *B*s, and (−) for all the *D*s and *F*s. (How the *C*s are handled varies from test to test.) You could then use these signs in one of many nonparametric tests for the median to determine if your students' grades were centered on a grade of *C* or not. **Rankings** order the measurements from one to the sample size of your dataset, with one being the largest (or smallest) value and *N* being the smallest (or largest). For example, rather than looking at letter grades in a final exam, you might look at the raw score. You could rank these scores from one to N, with one being the highest score. You could then use these ranks in one of several tests for the median to see if the median score was, say 75, or not.

Because most nonparametric tests rely on signs or ranks and not the original measurement values, they tend to be more robust than their parametric counterparts. For example, an outlier will receive the same sign whether it's one standard deviation above the median or six. In this way, an extreme value doesn't impact the outcome of a nonparametric test nearly as much as traditional tests that might use, for example, the sample mean. Also, because nonparametric tests are designed to apply to numeric and non-numeric data, most of them use the median rather than the sample mean to measure center location. After all, if you don't have numbers, you can't calculate an average. As long as you can order them, however, you can calculate the median.

There are many nonparametric methods available, more than I can cover in this book. In this chapter, I'll focus on three popular tests that illustrate how signs and ranks are used in the construction of nonparametric tests: the sign test, the Wilcoxon signed rank test, and the Kruskal–Wallis test for equality of the median.

The Sign Test

The sign test is a nonparametric hypothesis test for the median m of a population. Signs refer to the value of an observation relative to some reference value. If an observation is greater than the reference value, the sign is positive (+). If the observation is less than the reference value, the sign is negative (−). For example, say you have a set of observations and you'd like to know if the median of the underlying population is θ (the Greek letter theta). You could compare every observation in the sample to the value θ by assigning a sign to each observation, "+" for the greater-than-θ observations and "−" for the less-than-θ observations.

Now suppose you want to test whether the median of a population is θ or not. You could set up the following hypotheses:

$$H_0 : m = \theta \text{ vs.}$$

$$H_A : m \neq \theta$$

where m is the median. Because the alternative hypothesis has $m \neq \theta$, this is a two-sided hypothesis test. It doesn't need to be, however. You could set up a one-sided test by setting H_A to $m > \theta$ or $m < \theta$. For the sign test, the test statistic is the same. Only the critical values are different.

Recall that the median of a population is the 50th percentile, the halfway point in your sorted list of values. In other words, half of all the values in your population lie below the median and the other half lie above it. So if the null hypothesis is true and the median of the population really is θ, then roughly half of the observations should lie below the θ and the other half should lie above it. The sign test compares the fraction of observations below θ and above θ to determine if these fractions are sufficiently different from 50% to reject the null hypothesis that the median is θ. Here's how it works.

Take any observation at random. If the null hypothesis is true and the median of the underlying population truly is θ, the probability that observation will be greater than θ is 0.5. The total number of observations greater than θ in the entire sample follows the binomial distribution, that well-known probability distribution describing the number of successes in N independent trials. In other words, if you have N independent observations in your sample, each having a probability $p=0.5$ of being greater than the median, then the number of "+" signs, or greater-than-θ observations, follows the binomial distribution with number of trials N and success probability $p=0.5$.

From this point, implementing the sign test is as easy as calculating probabilities for a binomial distribution, which you can do with most basic data analysis software packages. Specifically, $\hat{n}$, the number of greater-than-θ observations is the test statistic. For the two-sided test, there are two critical values: the $\alpha/2$ and $1-\alpha/2$ critical value for the binomial distribution with N trials and success probability $p=0.5$. If $\hat{n}$ falls within these critical values, then H_0 is accepted. Otherwise, H_0 is rejected in favor of H_A.

There's one wrinkle in performing the sign test. In some cases, you may have observations that are exactly θ. These aren't positive or negative signs. The binomial test only allows for two possible outcomes (positive or negative). So, what should be done about these zero differences?

This is where the art of data analysis comes to play. Different statisticians have recommended different things. If you have a large sample with only one or two observations falling exactly on θ, then there isn't much harm in simply ignoring them and only using the positive and negative differences in your analysis (just remember to reduce your sample size accordingly). If you have a relatively small sample or many zero differences, then it pays to think about what these zero differences mean. A zero difference means an observation is exactly θ. This is strong evidence in support of H_0. The more of these zero differences you have, the stronger the evidence in favor of H_0. Some people have recommended flipping a coin and assigning and positive or negative difference value based on the outcome. However, because zero differences support H_0, I prefer to give them a value that more strongly supports H_0. This means allocating the zero difference observations in a way that closes the gap between the "+" and "−" groups. For example, suppose you have twenty observations, 6 positive signs, 12 negative signs, and 2 zero differences. The zero differences would be assigned to the "+" group,

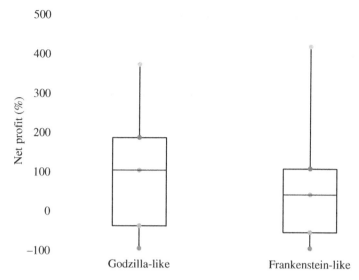

FIGURE 8.2   Net profit by movie monster-type.

because that would make the number of "+" signs and "−" signs more similar. If you had 8 negative signs, 9 positive signs, and 3 zero differences, two of the zero differences would go into the "−" group and one would go into the "+" group, giving ten "+" signs and ten "−" signs.

Say I'm a B movie producer, looking to make her first blockbuster hit. As a devoted monster movie fan, I decide to bring back one of my favorite classic monsters. But a good monster movie costs millions to produce, and I need to make money on this venture. Should I go with a giant creature who destroys cities without prejudice, something like Godzilla or King Kong, or should I go with a different type of monster, something more inhumanly human, something like zombies or Frankenstein? Which type of movie is more popular?

My studio has produced many monster movies over the years, so I go to box office profits and losses to answer this question. Figure 8.2 shows a boxplot of relative profits of the studio's monster movies broken out by type of monster. A negative profit means the movie lost money, whereas a positive profit means the movie made money. The profits range from −90%, for the stinker *Gorgon Goes Shopping* to 425% for the sleeper hit *Z-Wars*. Both categories of movies had big losses and big profits. Neither of these distributions looks symmetric and normal.

The median profit for the Godzilla-like movies is 108%. I can test whether or not a typical movie like this makes money by comparing the median profit to zero, the break even point. A one-sided sign test can do this for me. I add up the number of Godzilla-like movies with positive profits and get seven out of ten movies, or $\hat{n} = 7$. The $1 - \alpha = 0.95$ critical value for the binomial distribution with $p = 0.5$ and $N = 10$ is $n_{crit} = 8$. Since $\hat{n} < n_{crit}$, I'm forced to keep the null hypothesis that the median profit for Godzilla-like movies is zero. In other words, historically, there's no proof that this type of movie has been a money maker for my studio.

Repeating the process for the Frankenstein-like movies gives the following results. Ten out of sixteen of these types of movies have been money-makers, giving $\hat{n} = 10$. The $1 - \alpha = 0.95$ critical value for the binomial distribution with $p = 0.5$ and $N = 16$ is $n_{crit} = 11$. Since $\hat{n} < n_{crit}$, I cannot reject the null hypothesis that the median is zero. In other words, the median profit for Frankenstein-like movies is not significantly higher than zero either.

The Sign Test for Paired Data

The beauty of nonparametric statistical methods is that they are versatile. The sign test, for example, can be extended to paired data with almost no modification. Suppose you'd like to test the hypothesis that the median of two variables is the same. In other words,

$$H_0 : m_1 = m_2$$
$$H_A : m_1 \neq m_2$$

For numeric data, this test can be run simply by subtracting one variable from the other and plugging these differences into the original sign test for median $\theta = 0$. For an example of how this is done, see the case study at the end of this chapter.

Wilcoxon Signed Rank Test

The sign test is straightforward and uses no more sophisticated tools than those from the most basic statistics course. However, it's not the most powerful nonparametric median test out there. There are several alternatives, most of which can detect smaller deviations from the null hypothesis. These tests require more sophistication than the binomial distribution provides. The Wilcoxon signed rank test is just such a test.

The **Wilcoxon signed rank test** is a hypothesis test for the median of a population. Like the sign test, this test relies on signs. It's not as simple to run as the sign test, but there's a reason it's one of the most popular nonparametric tests available: it's powerful. Almost as powerful as the t-test when your data are normal, much more when they aren't. In other words, the Wilcoxon signed rank test works well under a variety of conditions, and so you can place a lot of confidence it its results.

The Wilcoxon signed rank test compares the following hypotheses:

$$H_0 : m = \theta \text{ vs.}$$

$$H_A : m \neq \theta.$$

where θ is the population median you'd like to test. Like the signs test, this test relies on the following argument. If θ truly is the median, in other words, if H_0 is true, then you can expect about half of your measurements values to fall below the median, and half of them to lie above it. So if you subtract θ from every observation to get differences d_i, then you'd expect about half of these differences to be less than zero and half to be greater than zero.

The Wilcoxon signed rank test uses the differences d_i by first sorting them by their absolute magnitude. In other words, the absolute values of the d_i are calculated and sorted from smallest to largest. The smallest absolute difference gets assigned a rank of one, the second smallest gets a rank of two, and so on. These ranks are then signed, meaning the original sign (+ or −) of the differences gets attached to its corresponding rank.

The test statistic for the Wilcoxon signed rank test is calculated directly from the signed ranks. The positive ranks are added and the negative ranks are added to get two values, $R+$ and $R-$. If the null hypothesis is true and your population median is θ, then $R+$ and $R-$ should be about the same. Why? Think about the process of constructing the signed ranks. As you travel down the list of ranked absolute differences, about half should have a positive sign and half should have a negative sign, and these signs should be distributed evenly throughout the list. So, not only should the number of positive and negative signed ranks be about the same, the sum of positive and negative signed ranks should also be about the same. Of course, since you're working with a sample and not the entire population, $R+$ and $R-$ won't be identical. The question is, are they different enough to declare statistical significance and reject the null hypothesis?

As with all hypothesis tests, statistical significance is determined by comparing a test statistic to a critical value. The test statistic is the smaller of the two sums, $R+$ and $R-$. Call this value R. For small sample sizes, say $N < 10$, the exact probability distribution of the test statistic should be used to calculate the critical threshold. A reference table for this critical threshold is provided in Appendix E. For $N \geq 10$, a normal approximation can be used to calculate the critical value. Specifically, take the test statistic R and transform it to

$$Z_R = \frac{R - (N(N+1)/4)}{\sqrt{N(N+1)(N+2)/24}}$$

Data value O_i	Difference $d_i = O_i - \theta$	Absolute difference $|d_i|$	SORT	Sorted absolute difference	Signed ranks	R+	R−
−90	−90	90		8.4	+1	1	
193	193	193		16	−2		2
8.4	8.4	8.4		36	+3	3	
181	181	181		82	−4		4
−16	−16	16		90	−5		5
378	378	378		181	+6	6	
−82	−82	82		189	+7	7	
287	287	287		193	+8	8	
36	36	36		287	+9	9	
189	189	189		278	+10	10	
					Total	**44**	**11**

FIGURE 8.3   Calculating the Wilcoxon signed rank test statistic.

Under the null hypothesis, the statistic Z_R is approximately standard normal. Therefore, you can compare it to the appropriate α critical value for the standard normal distribution. In either case, if the test statistic R is smaller than the critical value, then you reject the null hypothesis that the median is θ. Otherwise, you stick with the null hypothesis.

Many basic data analysis packages have the Wilcoxon signed rank test and these will do all the calculations for you. If yours doesn't, you can still use this test, just as long as you're willing to go through several data manipulations to find R. This process is illustrated in Figure 8.3 for median $\theta = 0$ (or $H_0: m = 0$) Godzilla-like movie profits. The sums are $R+ = 44$ and $R- = 11$. Since $R-$ is the smaller of the two values, the test statistic is $R = 11$. With $N \geq 10$, a normal approximation can be applied to get a test statistic of

$$Z_R = \frac{11 - (10(11)/4)}{\sqrt{10(11)(12)/24}} = -2.2$$

The 0.05 critical value for the standard normal distribution is $Z_{crit} = -1.64$. Since the test statistic is less than the critical value, $Z_R < Z_{crit}$, then we reject the null hypothesis and conclude that the median net profit for Godzilla-like movies is not zero.

When it comes to one-sided tests for the median, interpreting the Wilcoxon signed rank test results requires a little thought. $R+$ is the sum of signed ranks greater than zero, and $R-$ is the sum of signed ranks less than zero. The test uses the smaller of these two values, without regard for which alternative hypothesis is being used. If $R+ > R-$, then there's support for the hypothesis that the median is greater than θ. If $R- > R+$, there's support for the hypothesis that the median is less than θ. When the one-sided test rejects the null hypothesis, it doesn't tell you which of the alternatives is supported, only that one of them is. It's up to you to check $R+$ and $R-$ and determine whether it's the alternative you're testing for. For example, in the test for the Godzilla-like movie profits, $R+$ is the larger of the two rank sums. This lends evidence to the alternative that the median is greater than zero. As a result, you wouldn't reject the null hypothesis in favor of H_A: $m < 0$. You would, however, reject it in favor of the other alternative hypothesis, H_A: $m > 0$.

Like the sign test, the Wilcoxon signed rank test can be run on paired data just as easily as it can be run on simple data. Specifically, to test for equality of median, simply subtract the pairs to get differences and run the Wilcoxon signed rank test on the differences for $\theta = 0$.

Kruskal–Wallis

The sign test and the Wilcoxon signed rank test work smoothly for datasets containing a single variable or paired data. When you have more than two groups of data to compare, you need a nonparametric alternative to analysis of variance (ANOVA). In this case, the Kruskal–Wallis procedure is a great choice. The **Kruskal–Wallis** is a test for equality of medians between groups. Like the Wilcoxon signed rank test, this test uses ranks rather than the original observations. All observations, regardless of group, are ranked from smallest to largest. If the null hypothesis is true and the median of all groups is the same, then the sum of the ranks in every group should be about the same. The Kruskal–Wallis test calculates a test statistic based on the sums of ranks within each group and compares it to the critical value from the corresponding probability distribution. For large sample sizes, this probability distribution is the chi-squared distribution. For small sample sizes, a reference table based on the exact distribution should be used. Because the test statistic and the critical value are cumbersome to calculate by hand, for most applications, it's best to find a software package that offers the Kruskal–Wallis procedure and let it do the heavy lifting for you. As of this writing, Microsoft Excel does not offer a procedure for performing the Kruskal–Wallis test. Instructions on how to do this manually can be downloaded from *Kruskal–Wallis Test in Excel* (http://blog.excelmasterseries.com/2014/05/kruskal-wallis-test-alternative-for.html).

WHEN TO USE THE NUCLEAR OPTION

Because they don't make any limiting assumptions about your data, nonparametric methods can be used any time on virtually any dataset. This doesn't mean they *should* be used *all* the time, however. If your data conform to a known distribution, the appropriate parametric tests will have more power. In other words, it will be able to detect smaller deviations from the null hypothesis than the corresponding nonparametric test. This falls under the category of "if you have information, use it." If you know your data are approximately bell-shaped, then apply statistical methods that use this knowledge. Save the nonparametric tests for those difficult datasets you cannot easily characterize.

The following are some general guidelines about when to choose nonparametric tests over their parametric counterparts:

1. When your data clearly do not conform to a normal distribution and either, (i) you need to test for something besides the population mean or proportion, or (ii) you don't have enough samples to rely on the Central Limit Theorem.

2. When your data conform to a complicated distribution but you don't want complicated hypothesis tests tailored to this difficult, specialized distribution.

3. When your data are rankings, orderings, or non-numeric observations.

4. When you have potential outliers that might skew the results of traditional tests.

GODZILLA VERSUS KING KONG, ROUND TWO

The Internet is a great place to find opinions. And the top ten list seems to be one of the most popular forms of opinion out there. A recent search for "top ten movie monster list" produced millions of hits. Some of the hits were lists focused on slasher movies, ranking notorious killers like Freddy Krueger and Jason among the top contenders. Other lists tended toward modern monsters such as Predator and Cloverfield. King Kong and Godzilla are classic movie monsters, and so classic movie monster lists were the focus of my study.

(a)

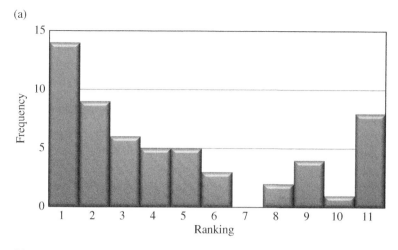

(b)

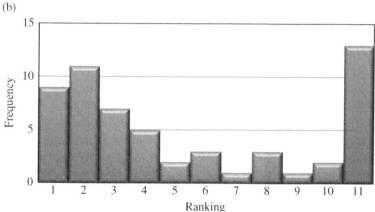

FIGURE 8.4 (a) Frequency distribution of Godzilla rankings (b) frequency distribution of King Kong rankings.

For complete details of my data collection process, see Chapter 7 of *The Art of Data Analysis: How to Answer Almost Any Question Using Basic Statistics*. To summarize, I grabbed a random sample of fifty-six top ten classic movie monster lists. On each list, the movie fan ranked his or her favorite monsters from one to ten, with one being the best. Both Godzilla and King Kong appeared on most of the lists, but not all of them. When one of the monsters didn't appear on a list, I used a common strategy for dealing with what are called **censored** observations by giving the monster a ranking of 11.

Figure 8.4 shows the histogram of Godzilla and King Kong rankings from my Internet sample. Comparing these histograms visually, it appears that Godzilla tends to have higher rankings (lower numbers). For example, the mode, or most common Godzilla ranking is 1. The mode of King Kong rankings, excluding the censored values, is 2. Only eight of the lists left Godzilla out of the top ten, while thirteen left King Kong out of the top ten. All of this tends to point to Godzilla as being more popular. Unfortunately, the difference in rankings wasn't big enough to declare statistical significance using the common parametric tests. In other words, Round one ended in a tie.

Certain nonparametric tests, particularly the Wilcoxon ranked sum test, have more power than the traditional tests when it comes to rankings and other oddball data. What will these tests show?

These data are paired. On every list, there's a ranking for Godzilla and a corresponding ranking for Kong, and the two values can be compared directly. In making the comparison, I subtracted the Kong rankings from the Godzilla rankings and performed the sign test and the Wilcoxon signed rank test for

$$H_0 : m = 0 \text{ vs.}$$
$$H_A : m \neq 0.$$

Results of the tests are shown in Figure 8.5. Unfortunately, Round two doesn't show anything different than round one did. Both tests recommend accepting the null hypothesis. In other words, the difference in rankings between the two monsters can be explained by random chance alone.

Disappointing? Yes. Surprising? Not really. With $N = 56$ observations, this is a large sample. The Central Limit Theorem promises that with a large N, even though the movie monster rankings aren't normal, the average difference as well as proportions calculated from the rankings are nearly so. In other words, the normal approximation is appropriate for these data, and so the t-test is more powerful than the nonparametric tests applied in this chapter. Since the t-test wasn't powerful enough to show a difference, then neither are the sign test and the Wilcoxon signed rank test.

And so, the battle of the movie monsters rages on.

Sign test	Wilcoxson signed rank sum test
Test statistic: –1.07	Test statistic: –1.56
Critical Values: –1.96 and 1.96	Critical values: –1.96
Conclusion: Accept H_0	*Conclusion: Accept* H_0

FIGURE 8.5 Godzilla vs. King Kong results for the two-sided sign test and Wilcoxon signed rank sum test.

BIBLIOGRAPHY

Hollander M, Wolfe DA. *Nonparametric Statistical Methods*. 2nd ed. New York: John Wiley & Sons, Inc.; 1999.

Jarman KH. *The Art of Data Analysis: How to Answer Almost Any Question Using Basic Statistics*. New York: John Wiley & Sons, Inc.; 2013.

Excel Master Series Blog. 2010. Kruskal–Wallis Test in Excel. Available at http://blog.excelmasterseries.com/2010/09/kruskal-wallis-test-done-in-excel.html.

Sheskin DJ. *Handbook of Parametric and Nonparametric Statistical Procedures*. 3rd ed. Florida: CRC Press, Inc.; 2004.

Sprent P, Smeeton NC. *Applied Nonparametric Statistical Methods*. Boca Raton: CRC Press, Inc; 2001.

9

MODELS, MURPHY'S LAW, AND PUBLIC HUMILIATION: REGRESSION RULES TO LIVE BY

I'm often reminded that I'm human. Sometimes, this reminder comes to me in the form of small mishaps—coffee spilled on a white carpet, an insensitive remark made to a friend, spinach stuck between my teeth—things I can correct with little embarrassment to myself. All too often, however, this reminder isn't nearly so gentle. Sometimes it hits me like a brick to the forehead. Like when I'm standing in front of a room full of important people, declaring the brilliance of my statistical conclusions to anyone who will listen.

My first data analysis humiliation occurred when I was only a few weeks into my career as a research statistician. I was working for a biomedical research company, call it MajorMedicCorp. This company developed new devices for monitoring a patient's blood chemistry. This company wanted to measure certain blood toxins noninvasively, meaning without drawing any blood. We knew we could measure how much life-giving oxygen was in the blood, simply by shining light through a patient's finger and measuring what came out. My task was to use similar light measurements to predict the amount of this toxic substance.

I was fresh out of school with a Ph.D. and lots of fancy statistical techniques at my fingertips, and this was a linear regression problem. Linear regression is undergraduate stuff. By all accounts, this task should've been

Beyond Basic Statistics: Tips, Tricks, and Techniques Every Data Analyst Should Know,
First Edition. Kristin H. Jarman.
© 2015 John Wiley & Sons, Inc. Published 2015 by John Wiley & Sons, Inc.

a breeze for me. Yet somehow, I still managed to turn this simple data analysis into a public humiliation. How could I possibly do this? Read on and find out.

MURPHY'S LAW AND SIMPLE LINEAR REGRESSION: A REVIEW OF TWO CLASSICS

Anything that can go wrong, will go wrong.

This is Murphy's law, and it's just as true today as it was sixty years ago when engineer Edward Murphy was credited with the famous saying. And when it comes to running a large medical research company, Murphy's law can be a source of embarrassment, cost time and money, and even cause major projects to be shut down. Suppose MajorMedicCorp has been experiencing an increase in the number of Murphy's law related mishaps. Workplace accidents, computer crashes, lost shipments, all of these things seem to be on the rise. Mishaps cost money and so naturally, management wants to minimize them. They want to know the impact Murphy's law is having on the bottom line and they decide to use me, the statistician, to figure out exactly what that impact is.

The CEO drops by my office and tells me he wants me to calculate the cost of Murphy's law. He gives me access to reams of employee data kept by Human Resources (HR), and I start digging. The amount of data HR keeps on the employees of MajorMedicCorp is a little scary, but also very useful. For example, HR keeps a record of lost work hours due to major workplace mishaps—accidents, injuries, computer failures, major miscommunications—and it also keeps tables of how much money each employee costs the company on a per work hour basis. Average cost per work hour times the number of lost hours gives the total cost related to Murphy's law. Simple enough.

But as I dig a little deeper, I notice the company has been growing in recent years. Eight years ago, there were 327 employees at this location. Now there are 521. More workers means more workplace mishaps, so I'd expect the total number of hours lost to such mishaps to be higher now than five years ago, simply because more people are working for the company. In other words, the number of hours lost to Murphy's law *depends* on the number of employees working for the company at any given time.

As a well-trained data analyst, I know the first step in the modeling process is visualizing, or plotting the data. When it comes to looking at

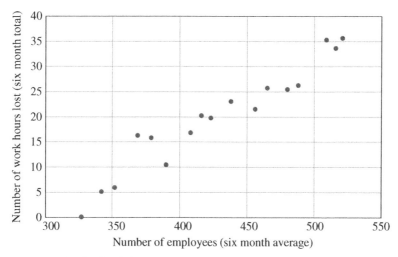

FIGURE 9.1   Work hours lost due to Murphy's law.

dependencies between variables, nothing beats a scatterplot. A scatterplot is an *x–y* plot, in this case, number of work hours lost to Murphy's law on the *y*-axis and the number of employees working for the company on the *x*-axis. Figure 9.1 shows a simulated version of this scatterplot, with the *x*-axis plotting a six-month average of the number of employees, and the *y*-axis plotting the total number of lost work hours over that same six-month period. These data fall along a nice, straight line. Simple linear regression ought to fit these data nicely.

Whenever you use a mathematical equation to predict the value of an unknown random quantity, you're using a **model**. Regression is one of the most widely used modeling techniques. A regression equation is a mathematical function that predicts the value of a continuous dependent variable, *Y*, as a function of one or more independent variables, X_1, X_2, and so on. Mathematically, a regression model can be represented like this:

$$Y = f(X_1, X_2, \ldots, X_N).$$

The simplest type of regression is **simple linear regression**, which uses a line with a single *X* variable, or

$$Y = mX + b.$$

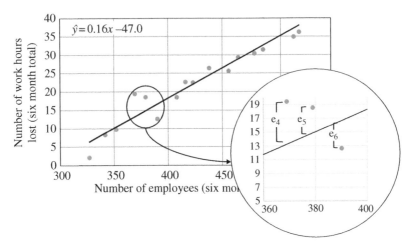

FIGURE 9.2   Simple linear regression on work hours lost to Murphy's law.

The values m and b are called regression coefficients. **Regression coefficients** are the unknown parameters in a regression equation, the values that relate X to Y. The purpose of regression is to estimate these regression coefficients. In a perfect world, if you picked the right regression equation for a dataset, it would be simple to calculate regression coefficients that took the X-values and predicted the Y-values with zero error. Of course, the world is far from perfect and every dataset has random variation in it. That's why we have least squares regression.

Least squares regression calculates regression coefficients. It does this by finding estimates for m and b, call them $\hat{m}$ and $\hat{b}$, that minimize the deviation between the observed Y-values and those predicted from the regression line. Figure 9.2 illustrates this deviation.

Mathematically, if $x_1, x_2, \ldots, x_N$ are the X-values in a sample, and $y_1, y_2, \ldots, y_N$ are the corresponding observed Y-values, then the error for the ith observation is the difference between the observed value y_i, and the estimated Y-value, $\hat{m}x_i + \hat{b}$, or $e_i = y_i - \left[\hat{m}x_i + \hat{b}\right]$. Least squares regression finds regression coefficients $\hat{m}$ and $\hat{b}$ that minimize the **total squared error**, or

$$\text{Total squared error} = e_1^2 + e_2^2 + \cdots + e_N^2.$$

The process of calculating the regression coefficients requires both calculus and linear algebra, so I won't go into the mathematical details of how this is done. Fortunately, you don't need these details to use linear regression. Most basic data analysis packages have a routine that will do it for you. The estimated regression coefficients for the Murphy's law data in

Figure 9.1, according to Microsoft Excel, are $\hat{m} = 0.16$ and $\hat{b} = -47.0$. Plugging these values into the simple linear regression equation gives an estimated Y-value, $\hat{y}$, of $\hat{y} = 0.16x - 47.0$.

Basic Regression Diagnostics

There are several statistics available for determining how well a regression line fits the data. Probably the most widely used diagnostic statistic is the R^2 value. The R^2 value measures the correlation between the observed y-values and the corresponding estimated y-values. The R^2 is a value between zero and one. An R^2 value near zero means the correlation is low and you have a poorly fit model. An R^2 value near one indicates a high correlation and a good fit to the data. Many data analysis programs report both the R^2 value and what's called an adjusted R^2 value. The **adjusted** R^2 value is generally a more reliable indicator of model accuracy, because this statistic takes into account the **degrees of freedom**, a reduction in the effective sample size caused by estimating multiple parameters from the same dataset.

The adjusted R^2 value is always smaller than the original R^2. For simple models, it's usually only slightly smaller, but for large, complex models, the difference can be noticeable. The original R^2 value for the regression model in Figure 9.1 is $R^2 = 0.93$. And because this is a small model with only two regression coefficients, the adjusted R^2 value is just slight less than this value at adjusted $R^2 = 0.92$.

While the R^2 value is a useful diagnostic tool, it's not without its problems. This statistic is notoriously impacted not only by how closely the y-values fall along the regression line but also on outliers as well as the range of x-values included in the data. A single outlier, especially at one end of the x-values, can shift a regression line, reducing overall accuracy while at the same time inflating the R^2 value. A small range of x-values can decrease R^2, even if the model is a good one. This is why any regression analysis should incorporate more than one diagnostic. A **residuals analysis**, an assessment of the residual error, is particularly useful, and like the R^2 value, it can be used for any regression model, simple linear or otherwise.

Residuals analysis usually starts with a residuals plot. The **residuals plot** is a plot of the regression error, the e_i values, in other words, the observed minus estimated y-values. When you have a well-fit model, the residuals plot should look like a bunch of points randomly bouncing around the line $y = 0$. Significant deviations from this pattern indicate problems. For example, outliers stand out as extreme values. A trend or systematic pattern in the residuals suggests the model needs more dependent variables. Typical

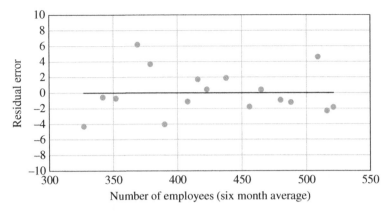

FIGURE 9.3   Residuals plot for simple linear model of work hours lost due to Murphy's Law.

patterns to look for are provided in many basic statistics texts, and I refer you Chapter 9 of The Art of Data Analysis: How to Answer Almost Any Question Using Basic Statistics for more information on this topic.

Figure 9.3 shows a residuals plot of the simple linear regression model for hours lost to Murphy's law. No outliers or obvious patterns can be seen in these residuals, suggesting the model is adequate.

Most packaged regression procedures will output at least one of two standard diagnostic tests, these tests being an analysis of variance and a t-test for significance of the regression coefficients. The analysis of variance is a test for overall lack of fit. This test compares two types of variation: (i) the **regression sum of squares** to (ii) the **residual sum of squares**. The regression sum of squares is the total variation captured by the regression equation, in other words, the amount of change in Y-values captured by the regression equation. The residual sum of squares is what's left over, the random error not explained by the model. If the variation captured by the regression equation is significantly larger than the residual error, then the model is declared statistically significant and appropriate for the data. Otherwise, it's declared to be ill fitting. Analysis of variance for the Murphy's law model in Figure 9.2 produces a p-value less than 0.01, suggesting the regression equation explains a significant portion of the overall variation in the data.

The t-test for significance of regression coefficients is just what the name implies: a t-test. The hypotheses for this t-test look at the regression coefficients m and b. Specifically, for the slope parameter

$$H_0 : m = 0 \text{ vs.}$$
$$H_A : m \neq 0.$$

The details of this test are slightly more complicated that a basic t-test for the mean of a population, but the end result is the same. A test statistic and p-value are produced. If the p-value is below the specified significance level, say 0.05, then the regression coefficient is declared to be significantly different from zero. The p-values for the intercept b and slope m of the work hours lost to Murphy's law are both <0.001, suggesting both terms are statistically significant and important to the model.

Simple linear regression is the first step on the path to understanding more sophisticated techniques. Fortunately, with a basic understanding of the least squares process and a working knowledge of common regression diagnostics, extending your repertoire to allow for nonlinear X–Y relationships and more than one X variable is straightforward. The next section describes common regression models that can be fit using basic least squares and a typical data analysis software.

BEYOND SIMPLE LINEAR REGRESSION: COMMON MODELS FOR COMMON SITUATIONS

The simple linear regression model for time lost due to Murphy's law appears to work pretty well. The regression line travels right through the center of the data, and the variation about the line is small. The slope of this line, 0.16, tells me that a typical employee loses an average of 0.16 hours, about ten minutes, of productivity to some mishap every six months. But average workplace mishaps are only part of the Murphy's law equation. A researcher who sits at a computer all day is much less likely to have an accident than, say, someone in the shipping department who hauls boxes for a living. So, this ten-minute figure, while a decent average number, probably won't apply to an office worker or to someone in the shipping department. A more accurate regression model might be constructed by breaking down the total number of employees into job categories based on the type of work they do—physical labor for the folks who build, move, clean, and ship things, office/clerical for the people who sit at computers most of the day, and laboratory for those employees who spend most of their time in a research lab. To break down the problem in this way, I need to go beyond simple linear regression.

Multiple Linear Regression

Suppose I'm interested in constructing a regression model for hours lost to Murphy's law as a function of the number of employees in each category physical labor (X_1), office/clerical (X_2), and laboratory (X_3). Regression like this, with multiple *x*-variables and a single *y*-variable, is called **multiple regression**.

The simplest type of multiple regression, **multiple linear regression**, uses a line to predict *Y* as a function of all the *X*-variables. The form of the equation looks like this:

$$Y = m_0 + m_1 X_1 + m_2 X_2 + m_3 X_3 + \cdots$$

Just like simple linear regression, the goal of multiple linear regression is to find regression coefficients m_0, m_1, m_2, ... that make the predicted *y*, or $\hat{y}$, as accurate as possible. And just like simple linear regression, the most popular technique for calculating these coefficients is least squares, which minimizes the total squared error, or deviation between the actual and estimated *y*-values.

To better understand Murphy's law at work, I break down the six-month average number of employees into job categories. The number of all three categories of workers has been steadily growing in recent years, with some up and down fluctuations. This is illustrated in Figure 9.4.

Many basic data analysis software packages offer multiple regression. You simply provide the *X* and *Y* values and the function will do the rest. Plugging the Murphy's law data into my Excel multiple regression add-in

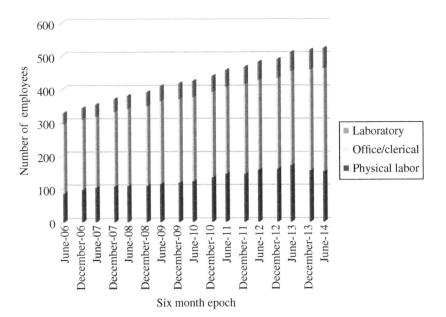

FIGURE 9.4   Average number of employees by job category.

gives regression coefficients $m_0 = -41.7$, $m_1 = 0.27$, $m_2 = 0.11$, and $m_3 = 0.05$. Since there are three X variables, it's impossible to generate a single Y vs. X scatterplot that illustrates the relationship between all of the independent and dependent variables. As a result, data analysts often use a scatterplot of estimated versus observed Y to evaluate their model. Figure 9.5a shows this for the work hours lost data.

A plot like this, along with the residuals plot in Figure 9.5b, is a great diagnostic tool for multiple linear regression. Ideally, the scatterplot will show points hugging the $Y = X$ line. Systematic deviations from this line suggest problems with the regression model. No such deviations are evident

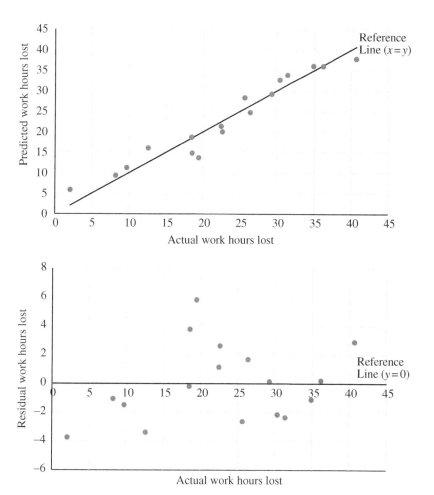

FIGURE 9.5   (a) Estimated versus observed lost work hours (b) and residuals plot for lost work hour.

in Figure 9.5, suggesting the multiple linear regression model adequately takes care of the dependencies between the X- and Y variables.

How does the multiple linear regression model compare to the simple linear regression model constructed in the previous section? The adjusted R^2 value for this model is 0.92. The adjusted R^2 model for the simple linear regression model is 0.92. The mean squared error of this model is 2.95. The mean squared error of the simple linear regression model is also 2.95. The residual plots for both models show no obvious patterns or model deficiencies. In other words, both models perform about the same. In this case, the only benefit in using multiple linear regression is the information I can get from the regression coefficients. The simple linear model only provides an average work hours lost per employee, any type of employee. The multiple linear model breaks this down for the three different categories of workers.

Polynomial Regression

Suppose the Murphy's law data looked like the data in Figure 9.6. Using a simple line to predict Y doesn't give very accurate results. The model underestimates the work hours lost in some places, and overestimates the work hours lost in others. In this case, the nonlinear relationship between the X and Y values requires something more sophisticated than simple or multiple linear regression.

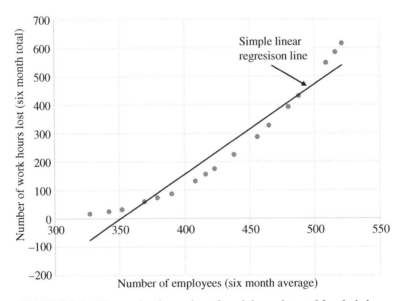

FIGURE 9.6   Not-so-simple version of work hours lost to Murphy's law.

Polynomial regression is a popular technique for taking nonlinearity into account. This technique is an extension of multiple regression where non-linear terms are added to the regression equation. For example, for a single X variable, the equation adds a term for X^2, X^3, and so on as needed:

$$Y = m_0 + m_1 X + m_2 X^2 + m_3 X^3 + \cdots$$

For multiple x-variables, polynomial regression adds squared terms (X_1^2, X_2^2, etc.), cubic terms (X_1^3, X_1^3, etc.), and mixed terms ($X_1 X_2, X_1 X_3$, etc.) as needed:

$$Y = m_0 + m_1 X_1 + m_2 X_2 + \cdots + m_{11} X_1^2 + m_{22} X_2^2 + \cdots + m_{12} X_1 X_2 + \cdots$$

Even though nonlinear terms have been added to the model, this is still considered linear regression *because the regression coefficients are all linear*. This is a good thing because it makes the regression process much easier. To perform polynomial regression, you simply create new variables in your data analysis software that correspond to the added terms ($X_1^2, X_1 X_2$, and so on) and then perform multiple linear regression.

Take the work hours lost data from Figure 9.6, for example. Suppose I'd like to construct a regression equation that takes the following form:

$$Y = m_0 + m_1 X + m_2 X^2.$$

In Excel, I'd add a new column containing X^2 and then perform multiple regression by including both X and X^2 terms. The results are shown in Figure 9.7. Clearly, this equation fits the Murphy's law data much better than the original simple linear regression model shown in Figure 9.6.

Polynomial regression is an incredibly versatile technique you can use to account for all kinds of nonlinearity in your data, especially if you already have some idea which terms should be included in the model. If you have no idea what the model should look like, this technique can still be very useful, however, deciding on which terms to include can be a challenge. Why? It's a mathematical truth that if you have $N(x, y)$ pairs, you can construct a $N-1$th-order polynomial that goes through every single one of the points. This law has serious ramifications when it comes to model building in statistics. In particular, given a sample of X and Y observations, you can construct a regression line that fits every one of the Y values perfectly if you only include enough polynomial terms in your model. This may sound like a good thing, but it isn't. A sample of data contains uncertainty and measurement error, so a regression model that estimates every single Y value perfectly is describing not only the dependent relationship between the X- and Y-values but also the

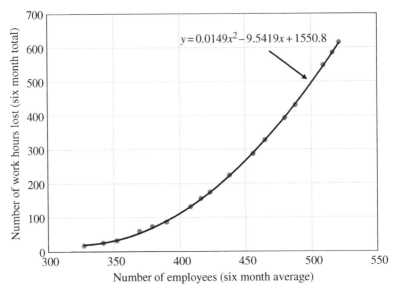

FIGURE 9.7   Polynomial regression and work hours lost due to Murphy's Law.

random measurement error. Such a model will have very high accuracy when it comes to the original sample, but very poor accuracy when it comes predictions. In other words, it models a sample of observations, but not the entire population that you're really interested in. This process of constructing a model with too many terms in it is called **overfitting**.

When it comes to polynomial regression, putting too many terms in the model, particularly mixed and high-order term such as X_1X_2, or X^3, can quickly lead to overfitting. Fortunately, there are strategies for minimizing the risk of overfitting. **Validation techniques**, discussed later in this chapter, divide a sample into subsets, so you can construct a model with some of the observations and estimate the accuracy of the model with the others. **Model selection techniques** are methods for selecting only the most statistically significant terms, and only including enough of them to adequately describe the X–Y dependencies in your data. One of the most popular model selection techniques is called stepwise regression.

Stepwise Regression

When it comes to constructing models, there's a prevailing philosophy among statisticians: simpler is better. This preference for simplicity, or **parsimony**, serves a couple of useful purposes. First, when it comes to explaining the results of an analysis, simpler models are much easier to understand and interpret than complex ones. Second, simpler models are harder to overfit to a dataset, particularly one with a limited number of samples. **Stepwise regression** is a technique for selecting terms in a multiple or polynomial regression model in such a way that only the most significant terms are included.

Recall that analysis of variance and the t-test are commonly used to determine whether or not a regression equation adequately describes the X- and Y-dependencies in a dataset. Specifically, the analysis of variance is a test for the overall significance of a particular model, while the t-test determines whether or not individual regression coefficients are significant. Stepwise regression starts with a basic model and then successively adds or removes terms based on the test statistics for these two tests. There are three types of stepwise regression:

1. Forward selection. Begin with the simplest possible model, containing only a constant term, and successively add terms until some stopping criterion is met.
2. Backward selection. Begin with the full model, the one containing all possible terms, and successively remove terms until some stopping criterion is met.
3. Combined forward/backward selection. Begin with the simplest possible model, and successively add and remove terms until some stopping criterion is met.

The three different approaches can easily produce three different regression models, so if your data analysis software package offers different stepwise regression options, it's worth running several of them so you can compare the differences.

Unfortunately, stepwise regression isn't always a cure for overfitting. All it can do is help you eliminate some of the unimportant variables or terms in your model. So, if you choose to use this method, it's important to apply one of the validation techniques described in the next section.

Multivariate Regression

For most everyday regression problems, the goal is to create a model for a single Y variable from one or more X variables. When you have more than one Y variable, you can usually construct a separate model for each Y variable based on the X variables. However, when you have Y variables that are correlated with one another, like they would be if, for example, they represent relative concentrations of chemicals in a mixture, it often makes sense to construct a single model for all of the Y variables at once. **Multivariate regression** is a technique for constructing a model that will simultaneously estimate multiple Y variables from multiple X variables. Multivariate regression is an advanced, less commonly used technique so I'll leave this topic to a more advanced textbook.

Nonlinear Regression

All of the methods presented here, even polynomial regression, fall under the umbrella of linear regression. This is because the models are linear in their coefficients. In other words, the regression coefficients are linear, for example, m_1 and not m_1^2, e^{-m_1}, $\sin(m_1)$ or some other complicated function of the original coefficient. When you have coefficients that lie inside nonlinear functions, you have a **nonlinear regression** problem.

Nonlinear regression models can't typically be solved using a simple one-step least squares procedure. They usually require iterative, more sophisticated techniques borrowed from a field of mathematics known as optimization. Many data analysis packages have nonlinear regression procedures in them. A detailed treatment of the topic is beyond this book, so I refer you to *Nonlinear Regression*, Seber and Wild (1989) for more information.

MISTAKES AND OTHER EMBARRASSMENTS TO AVOID

Lots of care should be put into constructing a statistical model, no matter how simple. All models have assumptions and limitations built into them, and in my experience, the moment you neglect the assumptions, forget about the limitations, or ignore a nagging suspicion about the applicability of a chosen statistical technique, that's when Murphy's law rears its ugly head. I've already mentioned two of the biggest potholes you can fall into when constructing a regression model. Underfitting leaves important terms out and doesn't capture all of the systematic X–Y dependencies in a dataset. Overfitting includes too many terms in the model, resulting in an overly optimistic estimate of the residual error. Unfortunately, these aren't the only problems to look out for. Linear regression makes several assumptions about a dataset, and not all of them always hold true. Two of the most important assumptions are as follows:

1. The Y values have random error as expressed in the E term of $Y=mX+b+E$. This error is assumed to be normally distributed with constant mean and variance.
2. There is no uncertainty in the X values. They are known with perfect accuracy.

The first assumption, that the measurement error of every Y value has the same variance, doesn't always hold true. Nonconstant variance often happens when the Y values are counts, areas, distances, or waiting times. With such measurements, the variance of the E term often increases as the measurement value increases. While the formal, and formidable, name for this phenomenon is **heteroscedasticity**, I'll just call it nonconstant variance here. Nonconstant variance can dramatically impact the quality of a regression model. In particular, the extreme Y values, those with the largest variance, can act like outliers, even though they're perfectly legitimate data points. They can exert their influence on a model by shifting regression coefficients away from their true values, resulting in a poor overall fit to the data.

A residuals plot will typically reveal nonconstant variance in the form of cone-shaped residuals, tightly clustered at one end of the regression line and spread wide at the other. To fix this problem, data analysts often perform a data transformation on the Y values. A **variance stabilizing transformation** is a function that changes the Y values in a way that makes the variance constant. For example, when the residuals look like a cone, narrow at one end and wide at the other, a log transformation is often used, where each Y value is transformed with the equation

$$Z_i = \log_{10}(Y_i)$$

A regression model for Z as a function of the X values is then constructed, and the estimated Y values are calculated by reversing the transform, in other words, setting

$$Y_i = 10^{Z_i}.$$

The second assumption in basic regression, that the X values are known with perfect accuracy, can be a problem with scientific studies in particular. In studies like these, researchers often want to make measurements for some variable X, and use them to estimate measurements for another variable Y. In cases like this, it's impossible to know the X values perfectly. In other words, there is uncertainty in both the X and Y values. There's usually nothing that can be done about this, so many data analysts just steam ahead with a regression model anyway, but this uncertainty in the X values does have an impact on the accuracy of the model. And because this uncertainty is not accounted for in the diagnostic tests and summary statistics produced by a typical regression procedure, the standard methods for estimating the model error don't apply. In this case, data analysts typically calculate the error in the regression model empirically, strictly from the data and without the use of standard regression diagnostics. And they typically do this using a technique known as validation.

Validation is the process of using some subset of observations to build a regression model and using others to calculate the error. There are a number of validation strategies. The simplest, and arguably best, validation technique is external validation. With **external validation**, you divide a dataset into two subsets: a training set and a test set. You build the model using the training set. You evaluate the model accuracy using the test set. Because you're assessing the model using observations that weren't used to build the model, the risk of overfitting is minimized.

In external validation, the model accuracy is typically measured using the **root-mean-squared error (RMSE)**. Similar to a standard deviation, the RMSE is the average deviation between the observed Y values and those estimated from the model. Mathematically, the RMSE can be written as follows:

$$\text{RMSE} = \sqrt{\frac{\left(y_1 - \hat{y}_1\right)^2 + \left(y_2 - \hat{y}_2\right)^2 + \cdots + \left(y_N - \hat{y}_N\right)^2}{N}}.$$

Unfortunately, there aren't any hard and fast rules regarding the number of training and test samples needed for external validation. Personally, I like to keep the number of training samples at about three times the anticipated number of regression terms, and use the rest for the test set. However, with very small datasets, it may be necessary to reduce the size of the training set. Dividing the dataset into training and test samples is best done with random sampling to avoid bias creeping into the model construction process. Simple random sampling works well for many applications, though certain situations might call for some of the more sophisticated techniques discussed in Chapter 2. To see external validation in action, go to the case study at the end of this chapter.

Though simple and effective, external validation requires plenty of data. When you have a limited sample size, cross validation might be a better option. **Cross validation** is a more economical alternative to external validation. Rather than leaving a significant number of observations out of the

WHAT I DID   **155**

training set, cross-validation fits the model over and over again, only leaving out a small number each time. An error is calculated for those hold-outs (the test set), and this is repeated until all observations have been held out once. These error estimates are then combined into a single error estimate for the model. There are many different versions of cross validation, among them leave-one-out and twofold cross validation, and the calculations that go along with it can get complicated. I'll leave the details of this strategy to the growing number of books that cover this topic, including *Regression modeling strategies: With applications to linear models, logistic regression, and survival analysis* by Harrell (2010). If you'd like some practical experience, many data analysis packages have cross-validation techniques built into them and if yours is one of them, I urge you to grab a dataset and experiment with some of the different options.

WHAT COULD GO WRONG, DID GO WRONG

"Tell me if the level of toxin Y in the blood can be accurately predicted using measured variables X_1, X_2, and X_3".

MajorMedicCorp was considering investing millions of dollars to develop a new medical device that could measure toxin Y, and this is what my boss, the Director of Research, asked me to do. It was my first major task with the company, and I was eager to prove myself. I had a dataset and a computer with my favorite data analysis software on it. All I needed to do was take the data and construct a regression model for Y as a function of X_1, X_2, and X_3. I didn't have any mathematical relationships to work with, I only knew we didn't expect the relationship between the X variables and Y to be a nice, straight line. Figure 9.8 illustrates how the data looked.

WHAT I DID

Figure 9.8 shows a relationship between the X and Y variables, but the scatter in these scatterplots is so large, there's no obvious choice when it comes to a regression equation. As a result, I decided to try polynomial regression with squared terms (X_1^2, X_2^2, X_3^2) being the highest order.

On paper, the regression was a simple exercise. Load the data into the data analysis software, run a polynomial regression on Y versus X_1, X_2, and X_3, and look at the regression diagnostics. I'd done this many times in school, and the whole process took less than two hours. The resulting model took on a formidable form. It looked something like this:

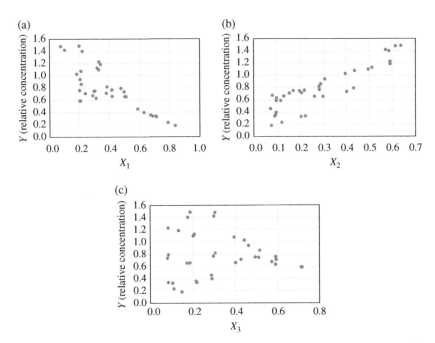

FIGURE 9.8   (a) Relative concentration of toxin (Y) versus measurement X_1, (b) relative concentration of toxin (Y) versus measurement X_2, and (c) relative concentration of toxin (Y) versus measurement X_3.

$$Y = 3.34 - 6.74X_1 - 3.42X_2 - 3.54X_3 + 3.17X_1^2 + 2.58X_2^2 + 0.47X_3^2$$
$$+ 4.57X_1X_2 + 5.59X_1X_3 + 2.59X_2X_3$$

Figure 9.9 shows a scatterplot of predicted versus observed Y value, a residuals plot, and three common summary statistics.

The adjusted R^2 was high, and the observed versus predicted Y values followed a reasonably straight line. The residuals showed a few extreme values, but nothing to give me concern. And the RMSE was 0.02, in other words, a mere 2% at the nominal concentration of one. The target RMSE for this measurement was 0.03 (3%) or better. A few short weeks into the job, and already I was a success.

I could've paused here, taking time to review my results and reflect on what I'd done. I could've, but I didn't. Instead, I rushed into my boss' office, anxious to show him what I found. He listened patiently, a smile growing on his face as I explained how we'd not only hit, but exceeded the target accuracy. Before he could give me more than an approving, "Looks good," we were interrupted by a visitor. It was the Vice President of the company.

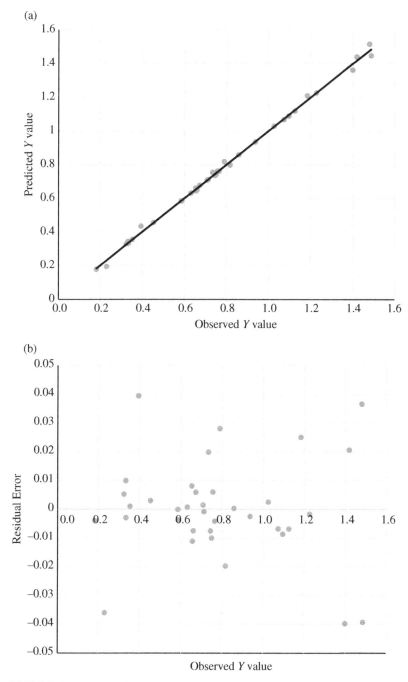

FIGURE 9.9   (a) Predicted versus observed Y values and (b) residual Y values

Apparently, he spotted the two of us talking and stopped by for a status update on this high profile project. He first addressed my boss, who reported that things were going well and I was fitting in nicely. Then, he addressed me. At this point, I could've taken my boss' lead and said something completely noncommittal, something like, "Yeah, things are going great." But I had these phenomenal results in my hand and I just had to share them. So I did.

The vice president was even more excited than my boss. He said he wanted me to brief the executive committee, so they could begin making plans to launch development of this new device. We set up a time for an initial briefing, and he rushed out the door.

Once he was gone, my boss turned to me. The smile on his face was gone. He said, "Before we brief the executive committee, I want you to run another dataset through your model and see how well it works." He handed me a disc with the new data and off I went.

The process took less than twenty minutes. I plugged the new X values into my regression model, and calculated the RMSE. Figure 9.10 shows the results.

About the best thing I could say about these results were that the predicted versus observed Y values still followed a roughly straight line. The scatter around that line was larger than I'd expected, the residuals showed a strange, indescribable pattern, and the RMSE was twice the value predicted from the regression model. In other words, my model might've worked well for the training data, but it was a failure when it came to prediction.

I rushed back into my boss' office, this time in a panic. I had to tell the vice president what I'd found and stop this runaway train from wrecking my career. Fortunately, my boss was a patient man and a good mentor for someone with more book knowledge than real-world experience. He offered to deal with the higher ups, postponing the executive briefing while I worked on figuring out what went wrong. Over the next few days, I gave myself a crash course in validation techniques and came up with a much better strategy for constructing this model.

What I Should've Done

After stewing over my dilemma and spending some time reading, I realized how the whole script should've played. First, I never should've said a word to the vice president about such preliminary results. He was a good manager, proactive and supportive, but cautious optimism wasn't in his repertoire. He wanted to take my results and act on them. Immediately. Second, I never should've rushed into my boss' office so soon after constructing the model. I should've taken a day to look closely at the residuals and review my data

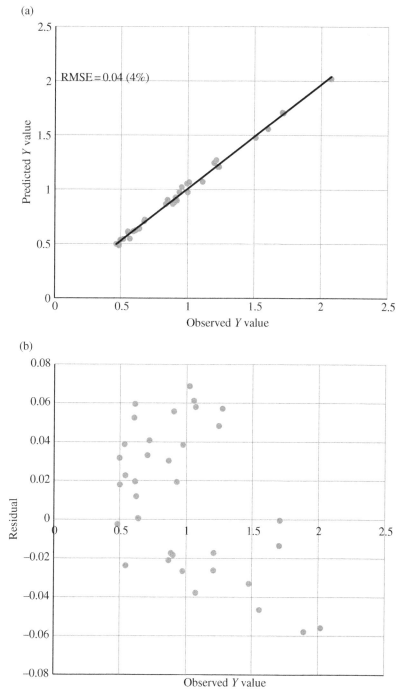

FIGURE 9.10   (a) Predicted versus observed Y values using a new dataset and (b) residuals for new dataset.

analysis. If I'd done this, I would've noticed a subtle bowed pattern in the residuals, and at the very least, this would've alerted me to some sort of model inadequacy. It might've even spared me the humiliation of having to admit I'd fit a bad model.

As far as the data analysis goes, I should've heeded all the warnings about the underlying assumptions made by linear regression. Not only did my Y values have measurement error, so did my X values. This uncertainty in the X values increased the scatter around the regression line, impacting the fit and causing diagnostic statistics like the adjusted R^2 value to be less informative than they otherwise would've been. I also should've been more concerned about including too many terms in the model. My model had ten terms. My sample had thirty-six observations. I assumed this was more than enough observations to eliminate the risk of overfitting. Sadly, it wasn't.

After a couple days, I constructed a completely new model. This is how the analysis went. The sample size was $N = 36$, and I split the dataset in half, choosing a test set of eighteen observations at random and setting them aside. Then, I performed stepwise regression on the remaining data, the training set. Doing this for the data in Figure 9.8 produces the following, much simpler model:

$$Y = 0.82 - 1.03X_1^2 + 1.68X_2^2 - 0.27X_3^2.$$

After constructing the model on the training set with acceptable results, I applied it to the test data. Figure 9.11 shows the results of the stepwise regression with external validation for the data in Figure 9.8.

The RMSE for this model is higher than the full model, but that's not a bad thing. The goal of regression is to model the underlying dependencies between the X- and Y-variables. There's a natural, inherent measurement error on top of this relationship and no amount of statistics can predict it. As a data analyst, your goal is to quantify that error, no matter what it is, so you aren't surprised when you apply the model to a new sample. This model does just that. The RMSE for the training data (0.05) is roughly the same as the RMSE of the test data (0.05). What's more, it's also close the RMSE of 0.055 for the new data set handed to me after the fact.

Like this model, my regression equation turned out to be very reliable. Unfortunately, the RMSE was higher than the target value of 3%.

Had I taken this approach in the first place, I never would've run into my boss' office declaring success. I never would've made the mistake of showing preliminary results to a vice president of the company. And I never would've had to stand in front of a room full of important people and explain why they shouldn't believe my results. Over the next few months, I redeemed

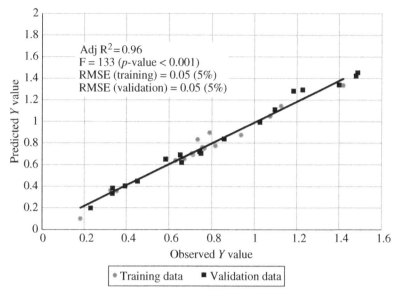

Adj $R^2 = 0.96$
F = 133 (*p*-value < 0.001)
RMSE (training) = 0.05 (5%)
RMSE (validation) = 0.05 (5%)

FIGURE 9.11 Predicted versus observed Y values using stepwise regression and an eighteen sample training set.

my credibility by coming up with a completely new model that did hit the target accuracy. I eventually left the company on good terms, with these rookie mistakes long forgotten by the fine people I worked with at Major MedicCorp. But these errors in judgement stayed with me, and even today, they serve as a constant reminder of what Murphy's law can do.

BIBLIOGRAPHY

Devijver PA, Kittler J. *Pattern Recognition: A Statistical Approach*. Englewood Cliffs: Prentice-Hall; 1982.

Geisser S. *Predictive Inference*. New York: Chapman & Hall; 1993.

Harrell FE. *Regression Modeling Strategies: With Applications to Linear Models, Logistic Regression, and Survival Analysis*. New York: Springer-Verlag, Inc.; 2010.

Jarman KH. *The Art of Data Analysis: How to Answer Almost Any Question Using Basic Statistics*. New York: John Wiley & Sons, Inc; 2013.

Montgomery DC, Peck E. *Introduction to Linear Regression Analysis*. New York: John Wiley & Sons, Inc; 1982.

Seber GAF, Wild CJ. *Nonlinear Regression*. New York: John Wiley & Sons, Inc; 1989.

APPENDIX A

CRITICAL VALUES FOR THE STANDARD NORMAL DISTRIBUTION

(a)

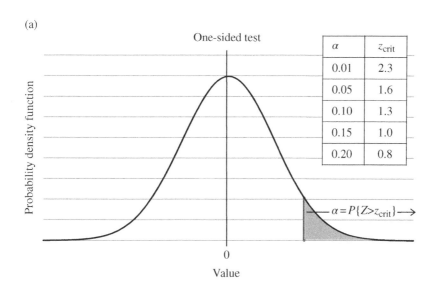

α	z_{crit}
0.01	2.3
0.05	1.6
0.10	1.3
0.15	1.0
0.20	0.8

One-sided test

Probability density function

$\alpha = P\{Z > z_{\text{crit}}\} \longrightarrow$

0

Value

Beyond Basic Statistics: Tips, Tricks, and Techniques Every Data Analyst Should Know,
First Edition. Kristin H. Jarman.
© 2015 John Wiley & Sons, Inc. Published 2015 by John Wiley & Sons, Inc.

(b)

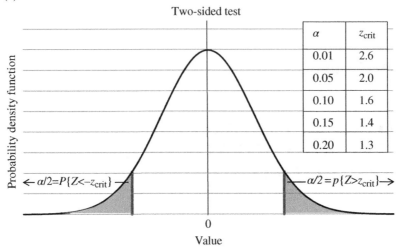

α	z_{crit}
0.01	2.6
0.05	2.0
0.10	1.6
0.15	1.4
0.20	1.3

APPENDIX B

CRITICAL VALUES FOR THE T-DISTRIBUTION

(a)

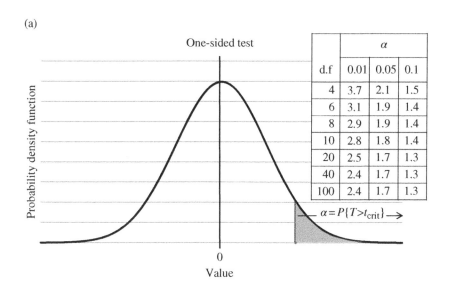

	α		
d.f	0.01	0.05	0.1
4	3.7	2.1	1.5
6	3.1	1.9	1.4
8	2.9	1.9	1.4
10	2.8	1.8	1.4
20	2.5	1.7	1.3
40	2.4	1.7	1.3
100	2.4	1.7	1.3

Beyond Basic Statistics: Tips, Tricks, and Techniques Every Data Analyst Should Know,
First Edition. Kristin H. Jarman.
© 2015 John Wiley & Sons, Inc. Published 2015 by John Wiley & Sons, Inc.

(b)

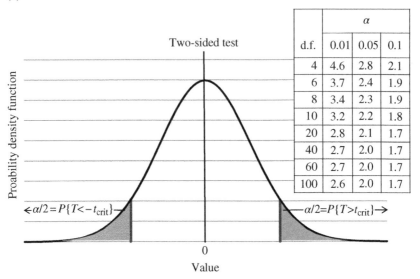

d.f.	α		
	0.01	0.05	0.1
4	4.6	2.8	2.1
6	3.7	2.4	1.9
8	3.4	2.3	1.9
10	3.2	2.2	1.8
20	2.8	2.1	1.7
40	2.7	2.0	1.7
60	2.7	2.0	1.7
100	2.6	2.0	1.7

APPENDIX C

CRITICAL VALUES FOR THE CHI-SQUARED DISTRIBUTION

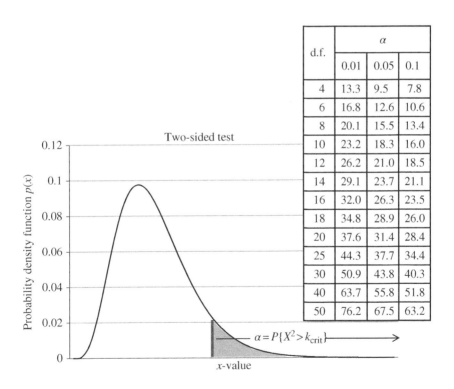

d.f.	α		
	0.01	0.05	0.1
4	13.3	9.5	7.8
6	16.8	12.6	10.6
8	20.1	15.5	13.4
10	23.2	18.3	16.0
12	26.2	21.0	18.5
14	29.1	23.7	21.1
16	32.0	26.3	23.5
18	34.8	28.9	26.0
20	37.6	31.4	28.4
25	44.3	37.7	34.4
30	50.9	43.8	40.3
40	63.7	55.8	51.8
50	76.2	67.5	63.2

Beyond Basic Statistics: Tips, Tricks, and Techniques Every Data Analyst Should Know,
First Edition. Kristin H. Jarman.
© 2015 John Wiley & Sons, Inc. Published 2015 by John Wiley & Sons, Inc.

APPENDIX D

CRITICAL VALUES FOR GRUBBS' TEST

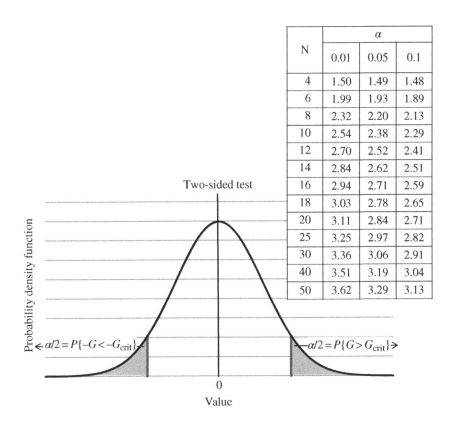

N	α		
	0.01	0.05	0.1
4	1.50	1.49	1.48
6	1.99	1.93	1.89
8	2.32	2.20	2.13
10	2.54	2.38	2.29
12	2.70	2.52	2.41
14	2.84	2.62	2.51
16	2.94	2.71	2.59
18	3.03	2.78	2.65
20	3.11	2.84	2.71
25	3.25	2.97	2.82
30	3.36	3.06	2.91
40	3.51	3.19	3.04
50	3.62	3.29	3.13

Two-sided test

$\alpha/2 = P\{-G < -G_{crit}\}$

$\alpha/2 = P\{G > G_{crit}\}$

Probability density function

0

Value

Beyond Basic Statistics: Tips, Tricks, and Techniques Every Data Analyst Should Know,
First Edition. Kristin H. Jarman.
© 2015 John Wiley & Sons, Inc. Published 2015 by John Wiley & Sons, Inc.

APPENDIX E

CRITICAL VALUES FOR WILCOXSON SIGNED RANK TEST: SMALL SAMPLE SIZES

(a)

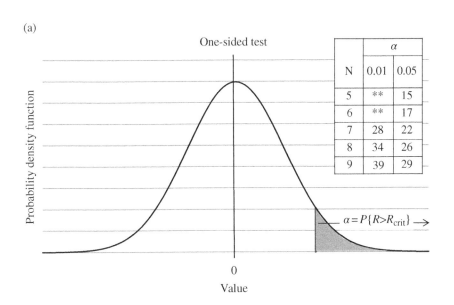

Beyond Basic Statistics: Tips, Tricks, and Techniques Every Data Analyst Should Know,
First Edition. Kristin H. Jarman.
© 2015 John Wiley & Sons, Inc. Published 2015 by John Wiley & Sons, Inc.

(b)

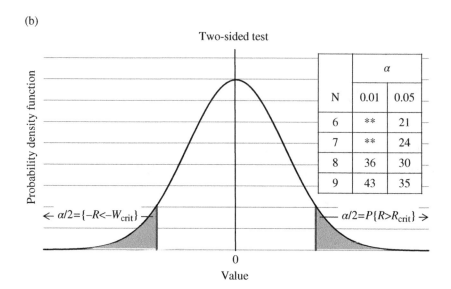

GLOSSARY

2 × 2 contingency table   a contingency table having one variable with two possible values and one outcome with two possible values.

2 × 5 contingency table   a contingency table having one variable with two possible values and one outcome with five possible values.

Alternative hypothesis   in a hypothesis test, the claim you accept as true only if you have enough evidence in the data to reject the null hypothesis.

Anderson–Darling test   a hypothesis test to determine if a sample conforms to a specific probability distribution, the normal distribution in particular.

Anomaly detection   the process of identifying outliers or other unusual observations.

Average   a measure of the central value in a set of observations. Also called a sample mean. Calculated by adding all the values and dividing by the sample size.

Bimodal distribution   a split frequency distribution, where observations are clustered around each of two central values.

Binomial distribution   a probability distribution describing the number of successes in a fixed set of independent trials.

Beyond Basic Statistics: Tips, Tricks, and Techniques Every Data Analyst Should Know,
First Edition. Kristin H. Jarman.
© 2015 John Wiley & Sons, Inc. Published 2015 by John Wiley & Sons, Inc.

Blocking in a study, the process of collecting data in groups, where each group is as homogeneous as possible; a sampling technique to minimize the impact of potentially confounding factors.

Bootstrapping a data-based method for assessing the accuracy of an estimate, whether it's the variance, bias, or a confidence interval. Based on resampling.

Boxplot A graphical representation of a five-number summary, illustrating the minimum, maximum, and quartiles.

Breakdown point the proportion of outliers a statistic can handle before it becomes impacted by these extreme values.

Censored observations observations known only to be less in some range of values. Observations whose exact value is unknown.

Chi-squared test a hypothesis test for determining if a sample conforms to a specific probability distribution. Applied to a contingency table, and can be used to test for independence between variables and outcomes.

Confidence interval a measure of confidence in a descriptive statistic. Also called the margin of error.

Confounding factor any variable that can impact the outcome of an experiment, making results ambiguous.

Consistent an estimate whose variance grows increasingly closer to the true value as the sample size grows.

Contingency table a tool for breaking down a frequency distribution, where frequencies are listed by variable and outcome with the variable values in rows, and the outcome values in columns.

Continuous observations observations that take on continuous values such as real numbers.

Continuous probability distribution a probability distribution for continuous observations.

Continuous random variable a variable, usually denoted by X or Y, that represents some as-yet-undetermined continuous outcome of a random experiment.

Control group in a study, a group of test subjects who are not subjected to a treatment. A group of subjects against which a treatment group is compared.

Controllable variable an independent variable that can be systematically manipulated in order to observe its impact on the outcome of a study.

Controlled experiment a data collection effort where variables or factors are tightly controlled.

Controlled trial a controlled experiment in which a treatment group is compared to a control group.

Convenience data a dataset that's easy to get. Data gathered out of seven convenience rather than from a designed study.

Correlation a relationship between two random variables or observations where the value of one affects the probability of the outcome of the other.

Critical value in a hypothesis test or confidence interval, the value of an observation or test statistic needed to achieve some small error probability.

Cross-validation a technique for estimating the accuracy of a model. Based on a process of successively removing observations from the sample, building a model, and calculating the accuracy of the left-out observations.

Cumulative distribution function (cdf) the probability a random variable is less than some value, $P\{X<=x\}$.

D'Agostino Pearson test a hypothesis test for determining if a sample conforms to a normal distribution.

Data transformation a function applied to a set of data values, usually to make the values conform to normal or other common probability distribution.

Decision criteria used in hypothesis testing, a rule for accepting or rejecting a hypothesis; usually, a threshold above which the null hypothesis is rejected.

Degrees of freedom the effective number of independent observations for a statistic, typically the number of observations minus one less the number of estimates.

Dependent variable a variable whose value depends on some other variable. In a study, the dependent variable is the outcome you'd like to measure. In regression, the dependent variable, Y, is the value you'd like to predict.

Descriptive research research based mostly on observational studies. Designed to characterize rather than establish cause and effect.

Descriptive statistics summary values calculated from a sample, designed to describe the center location, shape, texture, and other properties of a population.

Design of experiments (DOEs) established strategies for setting up different types of controlled experiments.

Discrete observations observations that can be counted, for example, whole numbers, counts, or categories.

Discrete probability distribution a probability distribution that describes discrete random variables.

Discrete uniform distribution a discrete probability distribution where the probability of an observation taking each possible value is the same.

Distribution-free methods Also called nonparametric methods. Statistical methods that do not depend on a specific probability distribution.

Double-blind study a study, typically involving a treatment and a placebo, in which neither the test subjects nor the experimenters know which subjects were given treatments and which were given placebos.

Effect in a study, it's the outcome or phenomenon you'd like to measure.

Effect size a method for measuring the practically meaningful difference between the H_0 mean and the actual mean.

Efficiency a comparison between the variance of a robust estimate and the sample mean.

Empirical probability a probability value calculated from data and not just a theoretical model.

Estimate a value calculated from a sample that estimates the corresponding value for an entire population.

Event a specific set of outcomes in a random experiment.

Expected frequency in a contingency table, the null hypothesis probability of each cell multiplied by the total number of samples.

Experimental design the science of planning experiments to produce data that will lead to clear, valid conclusions.

Experimental study a highly manipulated study. The independent variables are carefully controlled. The dependent variables are carefully measured.

Explanatory research research targeted to a specific goal. Typically involves controlled experiments.

Exploratory research the process of learning as much as possible about your subject matter. Typically includes background reading, convenience data collection and ad-hoc statistical analysis.

External validation a method for estimating the accuracy of a model. Performed dividing a dataset into a training set and a test set, building a model for the training set, and estimating accuracy for the test set.

Extreme value distribution a probability distribution that describes the largest (or smallest) value in a set of data.

Factor any variable that can impact the outcome of an experiment.

Factorial designs often used for experiments with a single dependent variable and many independent variables, each with a small number of possible values

False discovery   the process of finding statistically significant differences purely by chance.

Five-number summary   for a sample, the quartiles together with two more descriptive statistics: the minimum and maximum.

Five-point Likert scale   a system for gathering feedback on a survey. Based on a scale from one to five.

Fractional factorial design   an experimental design that takes a carefully chosen subset from the full factorial design and limits testing to those combinations.

Frequency distribution   a tally of the number of times each different category or value appears in a sample of discrete observations.

Full factorial design   a designed experiment where every possible combination of independent variables is tested.

Generalized extreme residual test   an outlier detection method where Grubbs' test is successively applied to a sample.

Goodness-of-fit test   a hypothesis test that compares a frequency distribution to some model probability distribution, with the goal of judging whether or not the data fit the model.

Grubbs' test   a formal hypothesis test for detecting outliers. Based on extreme values.

Heteroscedasticity   nonconstant variance in the observations of a sample.

High-order interactions   the impact created by combining many—usually more than two—independent variables.

Histogram   a method for graphing the frequency distribution of quantitative data. The data are binned, the number of observations in each bin are counted, and a bar graph of the counts is constructed.

Hypothesis test   compares an alternative hypothesis to a null hypothesis and determines if there's enough evidence to reject the null hypothesis.

Independent trials   successive random experiments where the outcomes are independent.

Independent variable   a variable or factor that does not depend on any other variable or factor.

Interactions   in a study, the impact of combinations of independent variables.

Interquartile range (IQR)   for a sample, the 75% percentile minus the 25% percentile.

Kolmogorov–Smirnov test   a hypothesis test for determining if a sample conforms to the normal distribution.

Kruskal–Wallis test   a hypothesis test for equality of medians between groups.

***K*th-order statistic**   the *kth* smallest observation in a sample.

Kurtosis   the fourth central moment. Measures the amount of bulge a probability distribution has.

Least squares regression   a method for estimating regression coefficients by minimizing the deviation between the measured *Y*-values and model predictions.

Levels   the possible values of a variable.

Linear regression   a method for relating two variables, *x* and *y*, through a linear function, for example $y = mx + b$.

Log transformation   a data transformation where the logarithm of each observation is calculated. Tends to make right-skewed data look more normal.

Median   a measure of the central value in a group of observations. The middle value.

Median absolute deviation   the median of the absolute differences between every data value and the median of the data values.

Mode   The most frequently observed observation.

Model   a mathematical function that uses a set of independent variables (*X*) to predict, describe, or classify a dependent variable (*Y*).

Model selection technique   in regression, a method for selecting a subset of independent variables (*X*) that adequately predict the dependent variable (*Y*).

Moments   parameters that specify the properties of a probability distribution.

Multiple linear regression   the process of constructing a line to predict the value of a dependent variable (*Y*) from more than one independent variables (*X*s).

Multiple regression   the process of constructing a linear or nonlinear function to predict the value of a dependent variable (*Y*) from more than one independent variables (*X*s).

Multivariate regression   the process of constructing a function to predict the value of multiple dependent variables (*Y*s) from more than one independent variables (*X*s).

Noncentral *t*-distribution   a *t*-distribution whose mean is not zero.

Nonlinear regression   the process of constructing a function to predict the value of a dependent variable (*Y*) from independent variables (*X*s), where the regression coefficients are raised to a power or imbedded in nonlinear functions.

Nonparametric method   a statistical technique that makes no assumptions about the underlying probability distribution of a sample.

Normal distribution a continuous probability distribution with a symmetric, bell-shaped curve.

Null hypothesis in hypothesis testing, the hypothesis assumed to be true until proven false with a statistically significant result.

Observations measurements, opinions, categories, or numerical values, anything that can make up a dataset.

Observational study a data collection effort where none of the variables or factors are controlled.

One-sided hypothesis test a hypothesis test where the alternative hypothesis looks for deviations from the null hypothesis in one direction only (either > or <). Critical values of one-sided hypothesis tests are α and $1-\alpha$, where α is the desired significance level.

Order statistics observations that have been sorted, or ordered, from smallest to largest.

Outlier an extreme observation, a value that sits far away from the majority of the observations.

Overfitting fitting too many terms to a statistical model, resulting in a model that incorporates not only the X–Y dependencies, but also random variation from the training sample.

Paired data observations collected in pairs that are not independent of one another.

Parameter a value, such as the mean or variance, that specifies key properties of a random variable or probability distribution.

Parametric test a test that assumes a sample conforms to a specific underlying probability distribution such as the normal distribution.

Parsimony the philosophy that all other things being equal, a smaller model is better.

Percentile the observation or measurement value below which a specified percent of the data fall.

Polynomial regression the process of constructing a polynomial function to predict the value of a dependent variable (Y) from independent variables (Xs), where one or more of the X values are raised to a power.

Population the collection of all people, places, or things under study.

Power of a hypothesis test, the probability of correctly rejecting the null hypothesis.

Practical significance in a hypothesis test, the smallest deviation from the null hypothesis that matters for a particular problem.

Prediction   in linear regression, the process of predicting y-values from corresponding x-values.

Probability   a number between zero and one that expresses how likely some future event is to occur.

Probability distribution   a mathematical formula for assigning probabilities to outcomes in a random experiment.

Q–Q plot   a plot of the data quantiles against the corresponding quantiles for a normal distribution. A graphical method for assessing the normality of a sample.

Qualitative data   observations that describe a category or type. Any measurement that cannot be sorted into a meaningful numerical order.

Quantiles   the x-values corresponding to incrementally increasing cumulative probabilities.

Quantitative data   numerical observations that can be sorted into a meaningful order.

Quartiles   the 25th, 50th, and 75th percentiles of a dataset.

Quota sampling   a sampling scheme where data are gathered until a prespecified number of samples in every subgroup of a population have been obtained.

Random experiment   a situation or trial where the outcome is not known beforehand.

Random sampling   choosing members of a dataset at random.

Random variable   a variable that represents the outcome of a random experiment.

Randomization   in a study, the process of randomly ordering the collection of data. A method for minimizing the impact of confounding factors in a study.

Range   the largest minus the smallest observation in a dataset.

Rankings   the rank, or order associated with every observation in a sorted dataset.

Regression   a method for predicting the value of a variable, Y, from another variable, X, using a mathematical function.

Regression coefficients   the values that relate independent variables (Xs) to the dependent variable (Y). The values that are estimated in the regression process.

Regression sum of squares   the amount of total variation in Y values captured by a regression equation.

Repeated measures study   a before-and-after study where test subjects are measured both before and after some treatment is given to them.

Replication   the process of collecting more than one observation in a study. A method for reducing the uncertainty in the results of a study.

Resampling   the process of sampling from a dataset, usually with replacement. The basis for using bootstrapping to estimate uncertainty.

Residual sum of squares   the amount of total variation in Y values not captured by a regression equation.

Residuals   the variation unaccounted for by a statistical model. In a regression, the actual y values minus the corresponding y values predicted by the regression line.

Residuals analysis   an assessment of the residual error in a regression model. Inspection of residuals to identify outliers and determine how well a regression line fits the individual y values.

Residuals plot   a plot of the regression error, or predicted minus actual Y values; tool for assessing the quality of a model.

Resistant statistics   estimates that aren't adversely impacted by extreme data, no matter how extreme that data may be.

Right-skewed data   a sample whose frequency distribution looks as if it's been stretched toward positive values, or to the right.

Robust statistics   estimates that are insensitive to outliers, skewed distributions and other nonlinear behavior.

Root-mean-squared error (RMSE)   the average deviation between the observed Y values and those estimated from the model. The square root of the sum of squared residual error, scaled by the sample size.

Sample   a carefully selected subset of the population. The data collected in a study.

Sample distribution   the probability distribution for an estimate, a statistic, or any other value calculated from data.

Sampling   the science of choosing a subset, or sample, for a study.

Sampling without replacement   the process of successively selecting members from a population without replacing them after they've been selected.

Scatterplot   an X–Y plot. A plot of two variables against one another. A way to visualize dependencies and relationships between two variables.

Shapiro–Wilk test   a test for normality. A hypothesis test for determining if a sample conforms to the normal distribution.

Significance level   the error probability specified by the data analyst when calculating confidence intervals or performing hypothesis tests. Usually denoted by α.

Signs (+ or −)   values assigned to each observation in a sample based on whether or not the observation is greater than (+) or less than (−) some reference value

Simple linear regression   a linear regression model with one independent variable (X) and one dependent variable (Y).

Simple random sampling   a sampling scheme in which every member of the population has an equal probability of being selected.

Simulated data   the process of generating data from a mathematical or statistical model or from some known probability distribution.

Skewed data   Data whose frequency distribution is shifted to the right or left of a bell-shaped distribution.

Standard deviation   the average deviation, or variation, of the observations in a sample around the center location.

Standard error   the standard deviation of an estimate, most commonly the sample mean.

Standard normal distribution   a special case of the normal distribution, where the mean is zero and the variance is one.

Standard normal random variable   a random variable that conforms to the normal distribution, has mean $\mu = 0$ and variance $\sigma^2 = 1$.

Statistically significant   a trend, pattern, or difference that's larger than expected based on random variation alone.

Stepwise regression   a technique for selecting terms in a multiple or polynomial regression model in such a way that only the most significant terms are included.

Study   a data collection exercise designed to answer some question about a group of people, places, or items.

Symmetric   a probability distribution or frequency distribution whose left and right halves are mirror images of one another.

Systematic sampling   in a study, the process of collecting data in some logical order.

Test statistic   a value calculated from the data. In hypothesis testing, a sample statistic used to accept or reject the null hypothesis.

Total squared error   in regression, the sum of squared residual error. A measure of the total variation unaccounted for by a regression model.

Treatment a primary independent variable in a study. Meant to change in the outcome of an experiment.

Treatment group the group of test subject who are subjected to the effect of an independent variable, or treatment.

Trimmed mean a robust version of the sample mean. Calculated by trimming, removing a percentage of the highest and lowest values, and calculating the average from the remaining values.

Two-sided hypothesis test a hypothesis test where the alternative hypothesis looks for deviations from the null hypothesis in any direction (usually $\neq$). Critical values of two-sided hypothesis tests are $\alpha/2$ or $1-\alpha/2$, where α is the desired significance level.

Two-way contingency tables a contingency table that breaks down a frequency distribution by one variable and one outcome only.

Type I error in a hypothesis test, rejecting H_0 when, in reality, H_0 is true.

Type I error probability in a hypothesis test, the probability of rejecting H_0 when, in reality, H_0 is true. The error α specified by the data analyst.

Type II error in a hypothesis test, accepting H_0 when, in reality, H_0 is false.

Type II error probability in a hypothesis test, the probability of accepting H_0 when, in reality, H_0 is false. Usually denoted by β and not controlled by the data analyst when performing a hypothesis test.

Unbiased estimate any statistic or estimate whose expected value is the same as the population parameter it's meant to estimate.

Uncontrollable variable an independent variable that cannot be systematically manipulated.

Validation technique a data-based method for constructing an accurate model. Involves dividing a sample into subsets, constructing a model with some of the observations, and estimating the accuracy of the model with the others.

Variance stabilizing transformation a function that transforms the observations of a sample into new observations with constant variance.

Wilcoxon signed rank test a hypothesis test for the median of a population. A powerful nonparametric alternative to the t-test.

Z-score an observation that has been scaled by subtracting the sample mean and dividing by the sample standard deviation.

INDEX

alternative hypothesis, 6, 40, 61, 127, 133
analysis of variance (ANOVA), 51, 93, 124, 144, 151
Anderson–Darling test, 102–3
anomaly detection, 109–22 *see also* outlier detection
ANOVA *see* analysis of variance
arithmetic mean *see* mean
average *see* sample mean

bell-shaped distribution, 2, 74–5, 76–8, 81, 84, 92, 113 *see also* normal distribution
Bernoulli distribution, 49–50, 52
bias, 72, 76, 80, 85, 104, 154
bimodal distribution, 124–5
binomial distribution, 50–53, 65, 81–2, 128, 130
binomial random variable, 82
blocking, 15–17, 25–6
Bonferroni adjustment,118

bootstrapping, 85–6, 88, 104
boxplot, 79–80, 88, 121, 129
breakdown point, 81–4

causation versus association, 14
censored observations, 135
central limit theorem, 51, 80, 93, 94, 116, 134, 136
central location, 83 *see also* descriptive statistics; median; mode; sample mean; trimmed mean
chi-squared distribution, 60–61, 63, 65, 67, 167
chi-squared test, 47–67 *see also* goodness-of-fit test
confidence interval, 5
  for a proportion, 1–2, 33–4, 37–8
  for the mean, 38–9, 95, 107–8
  robust, 81–2, 85–6, 88–9
confounding factor, 10, 13–17, 22, 26, 27, 29
consistent, 72, 76, 80, 92

Beyond Basic Statistics: Tips, Tricks, and Techniques Every Data Analyst Should Know,
First Edition. Kristin H. Jarman.
© 2015 John Wiley & Sons, Inc. Published 2015 by John Wiley & Sons, Inc.

contingency table, 55–8, 61–3
continuous data, 5, 48, 51, 53, 59, 63,
    72–4, 141
continuous probability distribution,
    92–6
continuous random variable, 92, 96
control group, 26–7
controllable variable, 21–2
controlled experiment, 21–2, 27
controlled trials, 21, 26–7
convenience data, 13, 18–20
correlation, 119, 143
critical value, 71, 116, 131
    for the binomial distribution,
        82, 130
    for the chi-squared distribution, 61,
        63, 65, 67, 167
    for a confidence interval, 33, 38,
        39, 41
    for Grubb's test, 116–17, 169
    for the Kruskal–Wallis test, 133
    for the normal distribution, 33,
        43, 163
    for the sign test, 127–8, 136
    for the t-distribution, 38, 39, 41,
        116–17, 165
    for Wilcoxon signed rank test, 132,
        136, 171
cross validation, 154–5
cumulative distribution function (cdf),
    92, 96

D'Agostino Pearson test, 101, 103
data transformation, 103–4 *see also*
    variance stabilizing
    transformation
decision threshold *see* critical value
degrees of freedom, 38, 41, 60, 63, 65,
    67, 117, 143
dependent variable, 6, 11, 22, 63, 141,
    143, 147
descriptive research, 21
descriptive statistics, 5, 71–5
    five-number summary, 79–80
    frequency distribution, 19, 53, 55, 57,
        58, 61, 67, 74, 81, 124–6
    interquartile range (IQR), 79, 84,
        114–15

median, 36, 79–83, 85–6, 88, 121,
    126–33
median absolute deviation
    (MAD), 84
mode, 54, 94–5, 136
sample mean, 4–6, 35–41, 51, 71–4,
    76–7, 80–81, 83, 85–6, 94–5,
    110–111, 113–14, 125, 127
standard deviation, 5, 36, 38, 40, 72,
    75–7, 84, 86–8, 95, 98, 110–111,
    113–14, 117, 125, 127, 154
trimmed mean, 82–4, 85, 86, 88–9
design of experiments (DOE), 22
    see also experimental design
discrete observations, 5, 47–67
discrete probability distribution,
    49–50, 92
discrete uniform distribution,
    59–61, 63
distribution *see* probability distribution
double blind study, 66

effect, 24
effect size, 40–41
empirical probability, 66, 102
error probability, 32
    significance and, 35, 117
    type I, 117
    type II, 40
estimate *see* confidence interval;
        descriptive statistics; regression
event, 48
Excel, 7, 97, 133–4, 143, 146, 149
expected frequency, 60, 62, 64, 67
experimental design *see* study design
explanatory research, 20–21
exploratory research, 17–19
external validation, 154, 160
extreme value distribution, 103, 116

factorial designs, 22–4
false discovery, 118
five number summary, 79–80
five-point Likert scale, 25, 53
fractional factorial design, 22–4
frequency distribution, 19, 53, 55, 57, 58,
    61, 67, 74, 81, 124–6
full factorial design, 22–4

generalized extreme residual test, 117
geometric distribution, 50
goodness-of-fit test 59 *see also* chi-squared test
  for continuous distributions, 101–2
  for discrete distributions, 59–65
Grubb's Test, 116–17, 121

heteroscedasticity, 153
high order interactions, 22, 24
histogram, 74–5, 94–5, 98–9, 104, 135
hypothesis test, 5–6, 35, 39–40, 71, 91–2
  analysis of variance (ANOVA), 51, 93, 124, 144, 151
  chi-squared, 47–67
  D'Agostino Pearson, 101, 103
  Grubb's test, 116–17, 121
  Kolmogorov–Smirnov test, 102–3, 106–7
  Kruskal–Wallis test, 136
  for the mean, 40–41
  nonparametric, 123–38
  normality tests, 99–100
  one-sided, 6, 40, 42, 43, 61, 127, 129, 133
  parametric, 124
  for a proportion, 42–3
  Shapiro–Wilk test, 102–3, 106–7
  sign test, 127–30, 136
  two-sided, 6, 40, 61, 127, 128, 136
  Wilcoxon signed rank test, 130–133

independence, 63–4
independent trials, 50, 52, 128
independent variable, 11, 14–16, 20–24, 27, 141
interactions, 22, 24
interquartile range (IQR), 79, 84, 114–15

Kolmogorov–Smirnov test, 102–3, 106–7
Kruskal–Wallis test, 136
kurtosis, 101

least squares regression *see* regression
levels, 22–3, 26

linear regression *see* regression
log transformation, 104, 106–8 *see also* data transformation

margin of error *see* confidence interval
maximum, 79–80, 121
mean
  of a probability distribution, 4, 50
  population, 4, 6, 35–7, 40, 71–2, 125, 134
  sample, 4–6, 35–41, 51, 71–4, 76–7, 80–81, 83, 85–6, 94–5, 110–111, 113–14, 125, 127
median, 36, 79–83, 85–6, 88, 121, 126–33
median absolute deviation, 84
mode, 54, 94–5, 136
model selection, 150
moments, 101, 103
multiple regression, 146, 149
multivariate regression, 151–2

negative binomial distribution, 50
noncentral *t*-distribution, 41
nonlinear regression, 152
nonparametric method, 102
nonparametric tests
  Kolmogorov–Smirnov, 102–3, 106–7
  Kruskal–Wallis, 136
  Shapiro–Wilk test, 102–3, 106–7
  sign test, 127–30, 136
  Wilcoxon rank sum test, 130–133
normal distribution 3, 33, 91–108, 112–13, 118, 132, 134, 163
null hypothesis, 6, 35, 39–41, 43, 59, 60, 65, 67, 102, 111, 116–17, 125, 131, 132

observational study, 14, 21, 27, 28
observations, 4–5
observed frequency, 59
one-sided hypothesis test, 6, 40, 42, 43, 61, 127, 129, 133
order statistics, 102

outlier detection, 109–22
  exact methods, 116–19
  rule of thumb methods, 111–16
outlier(s), 2, 53, 74, 76–7 *see also* outlier
    detection
overfitting, 150–152, 154, 160

paired data, 130, 133, 136
parameter, 4, 50, 71, 72, 142, 143
parametric test, 124
parsimony, 150
percentile(s), 79–81, 84, 86, 96, 115,
    121, 127
Poisson distribution, 50
polynomial regression, 148–50, 155
population, 4
power, 39–41, 101, 103, 117, 118,
    130, 134
practical significance, 35–7
prediction, 158
probability, 1, 3–5, 28, 32, 47, 51–2, 62,
    65, 66, 71, 81–2, 96, 101, 102, 104,
    106, 116, 118, 128
  distribution, 5, 6, 40, 49–50, 59, 60,
    92–4, 96, 131 *see also* continuous
    probability distribution; discrete
    probability distribution
  independent events and, 63–4
  p-value, 5, 6, 35, 39, 92, 101, 103,
    111, 125, 144, 161
  power, 39–41, 101, 103, 117, 118,
    130, 134
  significance level, 6, 35–7, 39, 41, 60,
    101, 103, 111, 117, 118, 131, 136,
    144, 145, 151
  type I, 39–40, 117
  type II, 39–40
probability distribution, 5, 6, 40, 49–50,
    59, 60, 92–4, 96, 131
  Bernoulli, 49–50, 52
  binomial, 50–53, 65, 81–2,
    128, 130
  chi-squared, 60–61, 63, 65, 67, 167
  continuous, 92–6
  cumulative, 92, 96
  discrete, 49–50, 92
  exponential, 103
  geometric, 50

negative binomial, 50
normal, 3, 33, 91–108, 112–13, 118,
    132, 134, 163
Poisson, 50
Student t-, 5, 38, 41, 65, 71, 116
proportion, 1–2, 33, 36–8, 42–3, 71, 92,
    103, 124, 134, 136

Q–Q plot, 95–100, 103, 106, 107
qualitative data, 51–53, 70
quantiles, 96–8, 102, 103
quantitative data, 51–3, 71
quota sampling, 29, 42

random experiment, 3, 4, 48–50, 52
randomization, 15–16, 18, 25, 26
random sample, 20, 29, 34, 71–2,
    76, 135
random variable, 3, 4, 49–52, 63, 82, 92,
    96, 112
range, 79
rankings, 126, 134–6
regression, 139–62
  coefficients, 142–4, 146, 148, 149,
    151, 152
  diagnostics, 143–5, 154, 155
  intercept, 145
  least squares, 142, 145, 152
  model, 141, 143–5, 147–9, 151
  multiple linear, 146, 149
  multivariate, 152
  nonlinear,152
  polynomial, 148–50, 155
  simple linear, 6, 140–145
  slope, 144, 145
  stepwise, 150–151, 160–161
regression sum of squares, 144
relative efficiency, 82–3
repeated measures study, 21, 27, 54
replication, 15–16, 18, 23, 24
resampling, 85–6, 88, 108
research methods
  descriptive, 21
  explanatory, 20–21
  exploratory, 17–18
residual analysis, 143
residuals, 143–4, 147, 153, 156,
    158–9

residuals plot, 143–4, 147, 153, 156
residual sum of squares, 144
resistant statistics, 78, 81, 86
right-skewed data, 75, 88, 94, 98, 99, 103–4
robust estimates, 69–90
  inter-quartile range (IQR), 79, 84, 114–15
  median, 36, 79–83, 85–6, 88, 121, 126–33
  median absolute deviation, 84
  mode, 54, 94–5, 136
  percentiles, 79–81, 84, 86, 96, 115, 121, 127
  trimmed mean, 82–6, 88–9

sample, 4–6, 20, 26, 28, 29, 32–4, 79, 85, 88, 103, 105, 106, 119, 131, 134, 135, 154, 160
sample mean, 4–6, 35–41, 51, 71–4, 76–7, 80–81, 83, 85–6, 94–5, 110–111, 113–14, 125, 127
sample size calculations, 31–44
sample standard deviation *see* standard deviation
sample statistic, 5, 6, 111
sampling without replacement, 33
scatterplot, 98, 141, 147, 156
Shapiro–Wilk test, 102–3, 106–7
significance level, 6, 35–7, 39, 41, 60, 101, 103, 111, 117, 118, 131, 136, 144, 145, 151
signs, 126–31
simple linear regression, 6, 140–145
simple random sampling, 28, 154
simulation, 110, 141
standard deviation, 5, 36, 38, 40, 72, 75–7, 84, 86–8, 95, 98, 110–111, 113–14, 117, 125, 127, 154 *see also* variance
standard error, 72, 82
standard normal distribution 33, 43, 93, 98, 113, 118, 132 *see also* normal distribution
standard normal random variable, 113 *see also* z-scores
standardized observations, 113

statistically significant, 35, 59, 61, 95, 122, 144, 150
stepwise regression, 150–151, 160–161
student *t*-distribution, 5, 38, 41, 65, 71, 116
study design
  blocking, 15–17, 25–6
  confounding factor, 10, 13–17, 22, 26, 27, 29
  controlled experiment, 21–2, 27
  controlled trials, 21, 26–7
  design of experiments (DOE), 22
  observational, 14, 21, 27, 28
  randomization, 15–16, 18, 25, 26
  repeated measures, 21, 27, 54
  replication, 15–16, 18, 23, 24
  survey, 21, 25–8, 53–5, 57, 58, 60–61
survey(s), 21, 25–8, 53–5, 57, 58, 60–61
symmetric distribution, 79, 84, 92, 100–101
systematic sampling, 28–9

t-distribution *see* student *t*-distribution
test statistic, 6, 39, 40
  for the Anderson–Darling test, 103, 106, 107
  chi-squared, 60, 62, 63, 65, 67
  for D'Agostino Pearson, 101, 103
  for Grubb's test, 116–17
  for the Kolmogorov–Smirnov test, 103, 106, 107
  for the Kruskal–Wallis test, 133
  for the Shapiro–Wilk test, 103, 106, 107
  for the signs test, 127–28, 136
  t-, 40–41, 145
  for the Wilcoxon signed rank test, 131–33, 136
  z-, 42–3
three-way contingency table, 57
total squared error, 142, 146
treatment, 26–7, 58
trimmed mean, 82–6, 88–9
t-statistic, 40–41, 145
t-test, 40–41, 51, 93, 111, 112, 124, 125, 130, 136, 144–5, 151

two-sided hypothesis test, 6, 40, 61, 127, 128, 136
2×2 contingency table, 56
2×5 contingency table, 56
two-way contingency tables, 56–7, 62–3
type I error probability, 39–40, 117, 118
type II error probability 39–40 *see also* power

unbiased estimate, 72, 76, 80
uncontrollable variable, 10, 14, 21–2, 27, 119
uniform distribution, 59–63, 103

validation, 150, 151, 154–5, 160–161
variance, 4, 36, 38, 50, 71, 72, 82, 85, 93, 101, 112–13, 153 *see also* standard deviation
variance stabilizing transformation, 153 *see also* data transformation

Wilcoxon signed rank test, 130–133

z-score, 98–9, 112–14, 116, 118, 121